Interactive
Packaging
Design

Interactive

Packaging

Design

The interactivity in packaging design can cause users' emotional interaction, thus generating the will to purchase, which is the significance of **interactive packaging design**.

DESIGN MEDIA PUBLISHING (UK) LIMITED

CONTENTS

004 **PREFACE**

006 **DESIGN INTERVIEW:**
INTERACTIVE PACKAGING DESIGN PROCESS

008 **CHAPTER 1**
WHAT'S INTERACTIVE PACKAGING?

008   I. Definition of Interactive Packaging

008   II. Origin and Background of Interactive Packaging Design

010   III. Types of Interactive Packaging Design

012   IV. Functions of Interactive Packaging Design

018 **CHAPTER 2**
DESIGN PRINCIPLES OF INTERACTIVE PACKAGING

018   I. Usability

018   II. Ease of Use

020   III. Agreeableness

022   Projects

064 **CHAPTER 3**
ESTABLISHMENT OF "INTERACTIVE RELATIONSHIP" IN INTERACTIVE PACKAGING DESIGN

064   I. Sensory Stimulation

066   II. Opening Way

068	III. Fusion of Feeling and Situation	
070	Projects	

118	**CHAPTER 4**
	DESIGN FACTORS IN INTERACTIVE PACKAGING DESIGN

118	I. Behaviour Factor
120	II. Humanity Factor
122	III. Visual Image
124	IV. Structure Design
126	V. Material Selection
128	Projects

208	**CHAPTER 5**
	TECHNOLOGY APPLICATION IN INTERACTIVE PACKAGING DESIGN

208	I. Application of Material's Sensitivity
208	II. Application of Material's Reaction Principle
209	III. Application of QR Code
210	IV. Application of Electric Components
211	V. Application of AR Technology
212	Projects

PREFACE

"Interaction" or "Interactive" refers to the information communication process between people and everything in the natural world in general and expresses the effect and influence between them. With the continuous development of science and technology, communication modes are changing correspondingly, and interaction is more and more emphasised. Today, more than ever, informatisation provides people broader interactive opportunities and ability to create interactive experience, bringing a brand new user experience.

For a long time, it is widely believed that packaging design is used to solve the issue concerning informational visual communication and belongs to graphic design category. As a subject, sometimes packaging design may also detach from the brand and exist independently counting on its novel design idea, unique design form and ground-breaking material technology, for relevant personnel to study. However, it is rarely seen that a case could solve marketing issue only by informational visual communication. A true lasting classic packaging work must represents its brand and combines various factors inlcuding market environment, consumption idea, brand background, product status, marketing planning, manufacture cost, etc. The packaging will run through the brand's life and infuse energy to it.

Strictly speaking, packaging design is a kind of interaction itself. As the carrier of a brand product, packaging will certainly interact and connect with users in the sales process. The brand owner always consider how to convey the product's properties and value points to users through visualisation. Interactive packaging design, subdivided from packaging design, places particular emphasis on how to increase interactive communication with users in form and considers more about the combination between users' consumption experience, product emotion and brand characteristics, and innovative application in form and technology. Around the brand's marketing mode, it will derive packaging characteristics more adherent to the brand's keynote.

With over 30 years of open development, China has become the second largest economic entity in the world. Along with rapid economic development and the rise of internet E-commerce platforms, China has formed our unique consumption culture and idea. Today, in face of consumption upgrade, the function of packaging is far from a "a silent salesman" which protects product transportation, storage and convey product information. The forms and functions of packaging design has become diversified and intelligent. More and more top brands have devoted into the development and application of interactive packaging.

On March 24th, 2017, Tmall launched a product called "Tmall Black Box". Many people thought it was packed in a black box. It turned out that it was "virtual packaging", which analyses users' preferences and through big data pushes different brands' new products according to their preferences. It was a new marketing platform for Tmall, an accurate promotion channel for brands, and a competent smart assistant for users as well.

In May 2017, on Tmall Super Brand Day, Mondelz China and Tmall launched a music box. 12 hours after launching, 20,000 (first batch) Oreo music boxes were sold out. The consumer could put a biscuit on the music box to play music and the music will change when you bite the biscuit and put it on the music box again. "Listening to music while eating" is achieved based on 5 weight sensors ringlike distributed in the put area. The system linked with put area will identify the amount of sensors that the biscuit covers automatically and the system will change music according to the amount change of covered sensors. In social network, Oreo developed its product advantages comprehensively and expanded topic discussion around its product to attract target consumer group to follow and forward. It expressed the brand's characters through the combination of product and packaging, thus creating a unparalleled brand. The interactive process between packaging and consumers presents the closest distance between the brand and its users.

In 2017, on FBIF, the largest food and beverage industry forum, chief designer of Motorola mentioned in his speaking a packaging project that he finished for Nutrilinx (see P244). The bottle cap could count the health care pills that was poured out and the relevant APP on smartphone will remind you which vitamins you should supplement today.

These cases remind us that the market trend of packaging design has changed immensely. Nowadays, with highly fragmented information, users have seen too much different communication forms, such as interesting illustrations, vivid posters, striking advertisement… However, besides these promotions, the products in their hands are nothing special. Some users even generate aesthetic fatigue after too much propaganda. The application of interactive packaging will break this phenomenon to some extent because it is no more just a concept demonstration on communication level but considers consumers' usage scenario and usage mode and practice their selection demand "from usage to emotion" of the brand and its packaging design.

ABOUT THE AUTHOR

Peng Chong, a rising Chinese designer of brand packaging, founder of Pesign Design, specially-appointed artistic design instructor of Guizhou University.

Peng Chong has won the platinum award in Pentawards (world top packaging design competition) with his work "QIAN'S GIFT", the highest award in single category of Greater China area. He has won multiple international design awards, including Pentawards, Red Dot, IF, Topaward Asia, World Star, etc. He has finished numerous successful commercial designs and served famous brands of China and abroad, including Fonterra, Changyu Group, Mengniu Dairy, Eastroc Beverage, New Hope Diary, etc. With a goal of "Design leads industry", he devotes himself to the exploration of emerging industry and development of eco materials.

DESIGN INTERVIEW

INTERACTIVE PACKAGING DESIGN PROCESS

Yeongkeun Jeong, born and raised in South Korea. A product designer and currently working as an industrial designer for IDEO(www.IDEO.com), Palo Alto.

1. Summarise the problem you set out to solve. What was the challenge posed to you? Did it get you excited and why?

The idea for "Butter Better" came to me on a picnic with friends. It was a really lovely day and we were sharing food, drink and each other's company when I realised that I had forgotten to pack a knife with which to spread the butter. Although eventually able to see the funny side of trying to spread butter with a floppy foil lid, I realised that this could be simply remedied with a small change to the lid of the packet. One product that combined both butter and knife would be functional and appealing. As I considered how to improve a long-standing and well known product, I became aware that modernising the product at the same time would enhance and change an everyday food item - bread and butter - into something much more special. A very simple picnic; bread and butter could be transformed into a special memory by including 'Butter Better' in a variety of colours and designs.

2. What point of view did you bring to the challenge? Was there anything additional that you wanted to achieve with this project or bring to this project that was not part of the original brief?

The most important factor I strived to achieve was designing a simple product that is both functional and convenient. I feel people respond positively to products which succeed in solving a problem they find annoying, especially if it is solved in an easy, simple and unexpected way. I wanted to adopt this logic to my design concept and I endeavored to find the ideal shape, colour combination, material and size to support my idea. In addition to the functionality of the product, I wanted this product to satisfy people both sentimentally and emotionally. This is a parallel to my belief that eating is not just a behaviour to achieve satiety, but involves all the senses. I set about to achieve this by thinking about how the product could appeal to various people. I believe I achieved this mainly due to the design: the materials used in the product, the way of representing logo, the colours etc.

3. When designing this project, whose interests did you consider? (Discuss various stakeholders, audiences, retailing, manufacturing, assembly, distribution, etc., for example.)

I focused mainly on how "Butter Better" would be interpreted by the consumer, so the audience was my primary concern. "Butter Better" needed to maximise ease of use as well as appealing aesthetically. The package needed to attract potential purchasers visually, suggesting that this product is essential for alfresco dining. Its small size and compact shape are important to consider; they need to be appropriate so it can be transported easily for lunches and picnics. The design of Butter Better also suits its function: being able to take off the lid to save packing two separate items. The second could be described as the 'point of purchase'. The package will be the first thing to be seen, therefore it is crucial for 'Butter Better' to be simple but attractive. This will encourage passers-by to pause, focus and appreciate the product. Each package fits when placed together in a specific way, the packets form a flower pattern. I wanted to link colours to particular flavours, for example, yellow colour represents honey butter and pink represents strawberry flavoured butter. I have considered this from the early stages as this will be useful for the manufacturers and distributers as it links into one simple package.

4. Describe the rigour that informed your design. (Research, ethnography, subject matter experts, materials exploration, technology, iteration, testing, etc., as applicable.)

I have spent much time designing a suitable shape and selecting appropriate material.

Firstly, I needed to think about the ideal shape which could satisfy both knife and butter container since these two need to be combined. I had to consider the knife tip which needs to reach the corner of the butter container. Additionally, I had to think about the angle and direction of the knife blade so that it is natural to use the inside of the knife for spreading after scooping. There is a preference for right handed people as they are the majority. I decided the final shape of the knife after experimenting with 11 differently shaped prototypes. There were two key materials available, that is plastic and wood. I asked some students in Material Engineering Science, University of Yonsei, practical questions like maleficence, sustainability and the cost of the materials. I also questioned 150 people about their feelings and emotions towards plastic and wood. Based on this information I opted for wooden material for the knife.

5. What is the social value of your design? (Gladdening, educational, economic, paradigm-shifting, sustainable, environmental, cultural, etc.) How does it earn its keep in the world?

"Butter Better" is a cheerful yet useful product that can be adapted to suit all ages and occasions. It provides a humourous answer to an age-old problem - where's the butter knife? By using biodegradable materials it is environmentally friendly whilst still addressing the social needs of providing a labour-saving. The tried and tested packets of butter are universal across the globe. This product will earn its keep by taking over that role.

6. If you could have done one thing differently with the project, what would you have changed?

My original design plan included an outer transparent plastic covering; however, I was unable to include this in my final design. It was not practical to have it in my final product, as I had insufficient money and also didn't have the correct equipment to add the outer wrap to the packets of 'Butter Better'. So I feel something wanting about it.

Chapter 1
What's Interactive Packaging?

I. Definition of Interactive Packaging

With the development of society and the increase of user experience, simple packages with only images and texts can no longer attract users and promote consumption. On this premise, interactive packaging emerges. It is a comprehensive design subject which integrates usability engineering, psychology, behaviour design, information technology, material technology and printing technology. Compared to traditional packaging design, the biggest difference is that interactive packaging design emphasises on the the interactive experience between users and packages. This interaction will lead users to pay more attention to packaging, thus achieving the goal to attract users.

See Picture 1.1.1 Work by Prompt Design. The drink comes with 4 flavours being the Orchid, Chrysanthemum, Lotus and Butterfly Pea. The designers wish this new package will give customers' experience and feeling of fresh and natural flower. So the bottle package is designed with a shrink film wrap (which is double-side printed) when torn along the dotted marks around the bottle top will show a blooming flower appearance with petals, bringing pleasant experience for users.

II. Origin and Background of Interactive Packaging Design

The development of packaging design reflects the development of human civilisation and technology.

1.1.1

Packaging originates from ancient times. People from primitive society used effective materials from natural world to pack items, such as kudzu vines, leaves, shells, animal skin, etc. The packaging at that time was mainly used for protection, storage and transportation. With certain development of productivity and technology, people began to use woven baskets and calcined vessels to contain items. At that time, packaging had presented formal beauty such as symmetry, balance, unity and variation. Besides utility function, packaging had aesthetic value too. With the emergence of the industrial age, more available packaging materials emerged, including glass bottles, metal cans, carton boxes, paper boxes, etc., which enriched the application of packaging design. Besides basic utility function and aesthetic value, packaging also acted as production instruction and attracted consumers.

In the 21st century, the range of goods is increased with our growing material and spiritual life, and consumption pattern has changed from sellers' market to buyers' market. Users' requirements on packaging are raised gradually: they desire "participation", "experience" and "feeling". In this background, interactive packaging design emerges quietly.

1.3.1

III. Types of Interactive Packaging Design

1. Function Packaging

Common function packaging such as antimicrobial and fresh-keeping packaging, anti-corrosion packaging is a scientific method to solve packing problems related to content. One type is to protect the content from interference effectively and keep it unstained. For example, vacuum packaging for food extracts all the air from the packing bag, which provides no survival condition for microorganism, to achieve the function of preservation and fresh-keeping. Another type is to eliminate certain defects and uncertainty for the product. For example, anti-corrosion packaging is to cope with the content's corrosion and is made of corrosion-resistant material.

See Picture 1.3.1 Work by Grantipo. If you can't improve a wine any more because its already perfect, the challenge is to keep it that way. That's why the designers created a 3.5mm thick stainless steel label that maintains a certain temperature which helps to conserve the qualities of the wine.

1.3.2

2. Sense Packaging

Sense packaging uses design methods (e.g. visual effect, smell, texture) to establish an external overall feeling for users. For example, a interesting visual image will cheer up and relax users' feelings and improve their impressions on the product. Some design combines the smell of food with its packaging and establishes a link between users and product through sense of smell.

See Picture 1.3.2 Work by Backbone Branding. The designers uses disposable cups as a means of promoting the restaurant brand. They draw faces on the cups and users could change the character's faces by rotating the outer packaging. The original idea is to change users' feelings through coffee and to provide them a positive attitude to face life.

3. Smart Packaging

Smart packaging refers to the packaging in which people use new materials or new technologies to transform the packaging structure or manage the product information integrally. It not only has general packaging functions but

1.3.3

also special properties. Now, smart packaging is divided into material-smart packaging, structure-smart packaging and information-smart packaging. It involves technical fields including machinery, electric, chemistry, etc. and mainly uses packaging materials' special properties such as photosensitivity, electro-sensitivity, humidity-sensitivity or air-sensitivity to identify key parameters inside the packaging, including illumination, temperature, humidity, pressure, etc. Or it tracks the product through internal sensing elements or advanced bar code and trademark information system, thus providing more accurate product data for users.

See Picture 1.3.3 Work by Great Advertising Group. It is a packaging for an album which uses photosensitive pigment to print. The print will emerged on the sun's rays only. Fans could listen through the app only if a smartphone was faced to the sunlight. In order to get the desired result, the designers made dozens of samples, since the photosensitive pigments by different brands in combination with paper and laminate behaved unpredictably, so they needed to find the most reliable option for print. The packaging was the essential part of the big promo campaign of the album. The idea of the packaging was welcomed and highly awarded by the famous design and advertising prizes. The album is out of stock now because of the high popularity.

IV. Functions of Interactive Packaging Design

1. To Highlight Brand Personality, Intensify Brand Positioning and Improve Competitive Advantage

A good packaging design is a communication bridge between product and users. Just like one's dressing style can reflect his/her personality, packaging can also reflect a brand's personality. Compared to ordinary packaging, interactive packaging is more humanised and it is easier to build a link between product and users. It conveys a high-end and friendly feeling which stimulates users to know the product and delivers product information to users clearly. Furthermore, it will enforce user's approval to the brand and improve his/her brand loyalty, thus forming a strong market competitiveness.

See Picture 1.4.1, Picture 1.4.2 Work by Lavernia & Cienfuegos. ZARA KIDS is a fragrance for children. Speaking to a child's visual language, the designers created two fun and attractive characters for boys and girls respectively. When rotating the lid, the eyes of each character change interestingly. This cute design can provide playful interaction for target customers, thus enhancing their impression and preference to the brand.

1.4.1

1.4.2

2. To Increase Product Packaging's Additional Value and Achieve Sustainable Use

With the development of commodity economy, the usage of packaging is increased. It is shown that packaging waste occupies a large portion in urban pollution. The sustainable use of packaging will reduce packaging pollution effectively. It refers to the extending function of packaging other than basic packing function. It requires the designer to expand his/her design thinking and consider packaging's extending function in a eco-friendly view. For example, a wine packaging could be used as pen holder or packing box could be used as a planting container.

See Picture 1.4.3, Picture 1.4.4 Work by Omdesign. First of all, this packaging uses green and eco-friendly material - cork. It consists of two cork blocks and an outer sleeve of pressed cardboard. Cork is 100% natural and sustainable, with special properties such as lightness, impermeability and thermal insulation, a typical Portuguese product. Secondly, this packaging enables consumers to use it creatively. It may be used afterwards for multiple purposes, for example as a pen holder, a flower vase or a sundrise box. This ensures that the packaging is not just used to transport and display the wine. The interactive packaging design endows it a higher use value.

3. To Improve Product's Reliability and Protect Users' Interest through Anti-counterfeiting Function

The new technology, new process and new material used in Interactive Packaging ensure its anti-counterfeiting function. It is more complicated and more difficult to replicate than ordinary packaging. For example, some interactive packagings use a structure design impossible to replicate to achieve a one-time packaging. Some integrate chips in the packaging to enable users to use related app to recognise the product through smart phones. This anti-counterfeiting function of interactive packaging can both protect manufacturers' and users' interests, and increase users' trust and sense of security to the product.

See Picture 1.4.5, Picture 1.4.6 Work by TO-GENKYO. False labeling on food is a worldwide concern. How to improve the label's anti-counterfeiting function to provide consumers safe food? TO-GENKYO designed a lable in the shape of an hourglass, which contains special ink that changes colour based on the amount of ammonia emitted by the meat, the longer the meat is stored, the more ammonia it releases. When the meat is no longer suitable for sale, the ink blocks the barcode at the bottom so that it cannot be scanned. This active visualisation of the product's shelf life creates a new relationship between consumers and comestibles.

1.4.5   1.4.6

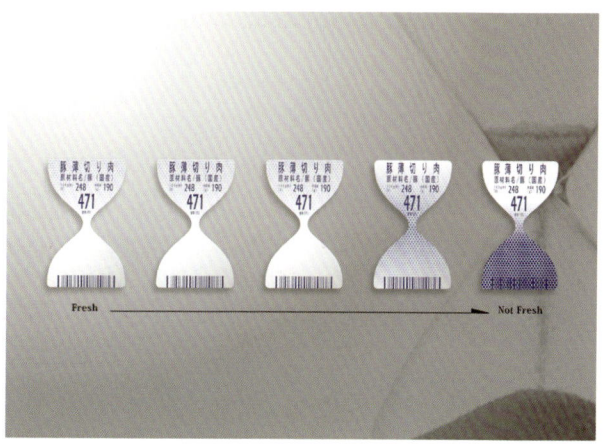

4. To Meet Users' Psychological Need and Improve Users' Satisfaction

After a period of quantity and quality need, people will have experience need for products. This change makes packaging design not only a practical science and also a practice of aesthetics and psychology. Today, in a high homogeneity commodity society, how to make your product stand out? The fun and interaction of interactive packaging can solve this problem. The designer should understand users' psychological need and stimulate their interactive experience through interesting interactive packaging, thus attract them to purchase the product.

See Picture 1.4.7, 1.4.8 Work by Grantipo. The designer hopes consumers could express their feelings and wishes through this packaging. The bottle, which is painted in a chalkboard-like material, comes equipped with sticks of chalk so that the consumers may personalise it in any way they please. If they're indecisive, it can be changed and modified at any time with the swipe of a towel. Consumers could participate in the packaging design. Not only they could express their feelings, but also create a unique personalised packaging.

1.4.7

1.4.8

Chapter 2
Design Principles of Interactive Packaging

2.1.1

I. Usability

Packaging plays an important role in product quality and a qualified packaging design must be usable. Usability is a basic requirement of interactive packaging design and the minimum requirement. Unusable packaging will be replaced and eliminated in the end. For example, most drinks are contained in glass bottles in earlier stage. However, glass is heavy and fragile, inconvenient for transportation and distribution, so it is replaced by plastic packing.

See Picture 2.1.1 Work by Brand Brothers. This is a packaging design for fertilizer. Besides product information, cutting marks, ruler and gardener's calendar are also printed on the packaging. When the pack is empty – a gardener can use it as a container for seedling. The "gardener's calendar" suggests making marks about seeding and nourishing. This additional function of the packaging gives it a "second life" and a higher usability.

II. Ease of Use

The ease of use of interactive packaging design is reflected in its practicability and convenience. The designer should consider in the users' view and refer to users' psychological needs, physiological habits and ergonomics to create a packaging that accords with users' habits. Ease of use is a higher design principle and is very important to improve users' recognition to the product.

See Picture 2.2.1 to Picture 2.2.3 Work by Yeongkeun Jeong. Whenever we eat bread, at the picnic, in the cafe or on an airplane, we usually use disposable butter. We need to spread the butter and this design saves you the trouble to find the tool. The designer replaced its ordinary container lid with a wooden, knife shaped one. This way butter can be easily and quickly spread. Users could choose butter flavour according to the colour of the wood knife. This container is not only easy and fast to use but also it makes daily routine of spreading butter more interesting and exciting.

See Picture 2.2.4 to Picture 2.2.6 Work by Backbone Branding. It is a package for high-end flower brand. In general, it looks like a bouquet in a big box. However, you will find that the whole packaging design is divided into 6 parts when you take off the cover. You can arrange the bouquet as you wish, to create different beautiful mood.

2.2.1

2.2.2

2.2.3

2.2.4 2.2.5 2.2.6

2.3.1

III. Agreeableness

Agreeableness of interactive packaging design emphasises on "people oriented" idea and is the highest level principle. An excellent interactive packaging design should start from users' emotional needs and combines emotions into packaging. It should meet users' spiritual needs through a good emotion communication between user and product, thus stimulating users' resonance.

See Picture 2.3.1 Work by BBDO. BECK'S is a traditional beer brand in Germany. However, more and more young beer drinkers tend to prefer small breweries and craft beers instead of big players like BECK'S. To attract these younger consumers, the designers turned the whole bottle into an aluminum canvas so that consumers may scratch favourable images on it. This interesting interaction attracted more younger audience for the brand. In consideration of young people's psychology for fun and interesting items, this design builds a link between user and product.

See Picture 2.3.2, Picture 2.3.3 Work by Art of Mind. This is a milkshake cup design and both the lip and cup are made of plastic. There are 9 different characters printed on the outer paper sleeve, and almost every character has perforated ears, which makes the cup characters more lovely and provides extra fun for users.

2.3.2

2.3.3

HANDEN PET CAN

Design agency:
www.xiangaopeng.com

Creative director:
Go Peng

Designer:
Go Peng

Photography:
Xu Yan Liang, Zhang Long Yuan

Country:
China

What's Unique?

The packaging design combines the tab with the tongue of the pet. While opening the can, customers achieve interaction with the packaging, implying the good taste of pet food.

The packaging design combines the aluminium foil of the can and takes the vector shape of pet head as the main graphics, which maximises the visual attack of the packaging. With the tab of the can replaced with tongue of the pet, the whole design highlights the creation while differentiates the brand from similar products, establishing a good base for the brand's marketing.

SIMAIYUN

Design agency:
www.xiangaopeng.com

Creative director:
Go Peng

Designer:
Go Peng

Photography:
Xu Yan Liang

Country:
China

Award:
Pentawards 2017 Nomination Award

What's Unique?

The packaging design of SIMAIYUN combines the chef character and the vermicelli product. When opening the package, customers will interact with the product, highlighting the concept of good taste.

The designer created two chef characters to present the good taste and they combined the design with the product's shape. The whole packaging design is integrated with handmade vermicelli to show the greatest selling points: nutritious and delicious.

SIHEYI

Design agency:
www.xiangaopeng.com

Creative director:
Go Peng

Designer:
Go Peng

Photography:
Xu Yan Liang, Zhang Long Yuan

Country:
China

What's Unique?

The packaging design of SIHEYI chopsticks combines the virtual characters and real chopsticks to create interesting interaction between the product and the packaging, creative, direct and unique.

With the rise of take-out market, young white-collar workers have become the main they consumer group of the market. Therefore, the designer created a group of restaurant waiter and waitress for the packaging of chopsticks. Part of the packing bag is made transparent so the chopsticks will become a part of the character. The whole packaging combines virtuality and reality, differentiating it from similar products.

    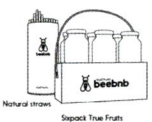 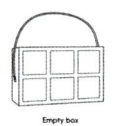 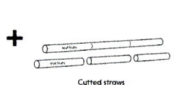

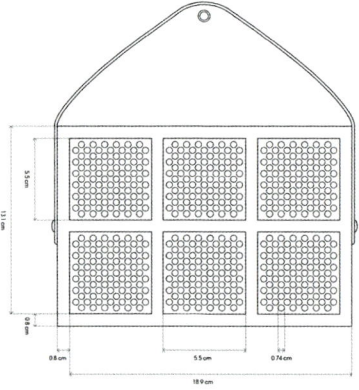 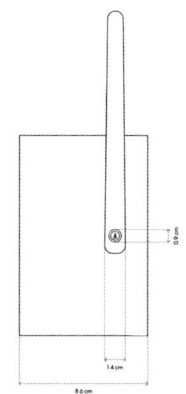 

BEEBNB

Design agency:
BBDO

Designer:
Nicolas Molès

Photography:
Christian H. Hasselbusch

Creative director:
Daniel Schweinzer, Lukas Liske

Country:
Germany

Award:
Art Directors Club für Deutschland: Bronze
Eurobest: Bronze
Comm Awards: Silver

What's Unique?

Beebnb is a wooden sixpack that turns into a bee hotel in four simple steps.
1. Enjoy your true fruits smoothies.
2. Cut the straws in three equal pieces.
3. Arrange them in the empty six-pack.
4. Install Beebnb in a sunny spot.

All materials used are natural and sustainable. Instead of being thrown away, they give bees a home. Beebnb is used for in-store promotions in organic food shops.

Big parts of the world's bee population are endangered. As a smoothie brand True Fruits is directly dependant on the work of the bees: no pollination, no fruits, no smoothies. That's why True Fruits was looking for a simple way to contribute to the preservation of bees.

BBDO came up with an upcycling design idea that turns a sixpack into a bee hotel: beebnb. It works by combining the wooden box with biological straws – creating a scientifically proven refuge for bees in four simple steps. So every True Fruits client can help to cultivate the next generation of bees. And thereby, tasty True Fruits smoothies.

CUATRO ALMAS/ CORK

Design agency:
Grantipo

Creative director:
Sergio Daniel García

Designer:
Sergio Daniel García

Client:
Bodegas Señorío de Somalo

Country:
Spain

What's Unique?

If this wine has a lot to say, you, too sure, can also arrange few pins to show what you want.

There is one thing that distinguishes the great wines of the low-end: with the cork. To increase the perceived quality of this wine, it also put the cork out, improving pass grip and warmth of the bottle.

INVITATION FOR REPRESENTATIVES OF THE PRESS

Design agency:
kissmiklos

Photography:
Eszter Sarah

Country:
Hungary

What's Unique?

The designer came up with a creative invitation, which is related to biking and Liget as well. He bought 14 pieces of green cold pressed juices and designed a typical Juice label, which was the invitation. They sent every final personalised juices with bike couriers to the given addresses.

The objective of the Liget Budapest Project is for the renewed City Park to become a tourist destination with a complexity, quality and international appeal unrivalled by any other in Europe. Before the construction, and after the winner landscape architectural plans, they organised a bicycle picnic, where they would like to reconciled about the new cycle ways.

GOOD HAIR DAY PASTA

Design agency:
Nikita

Designer:
Nikita Konkin

Country:
Russia

What's Unique?

What a good hair day for Pasta! Nikita creatively made use of the strands and shapes of pasta to create an interesting series of packaging that definitely capture attention on the shelves. An awesome head-turner!

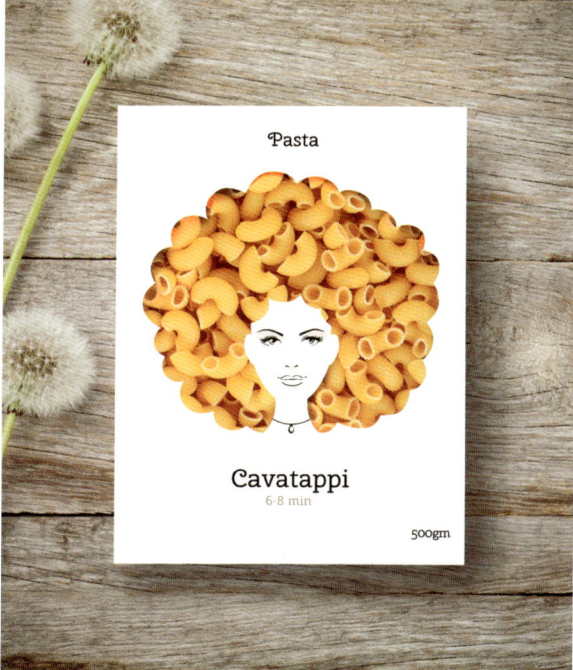

This brand of original Italian pasta, comes from the heart of the Italia region of Abruzzo. This packaging is internationally well-known designed by Nikita Konkin. It has won a number of prestigious international awards. Enjoy this masterpiece of food art and be on the Italian atmosphere.

YUMMY BITES 123

Design agency:
The 1984

Creative director:
Andreas Junus, Wanda Kamarga

Designer:
Tifannie Budiono, Fran Hakim

Client:
CV Kusuma Food

Country:
Indonesia

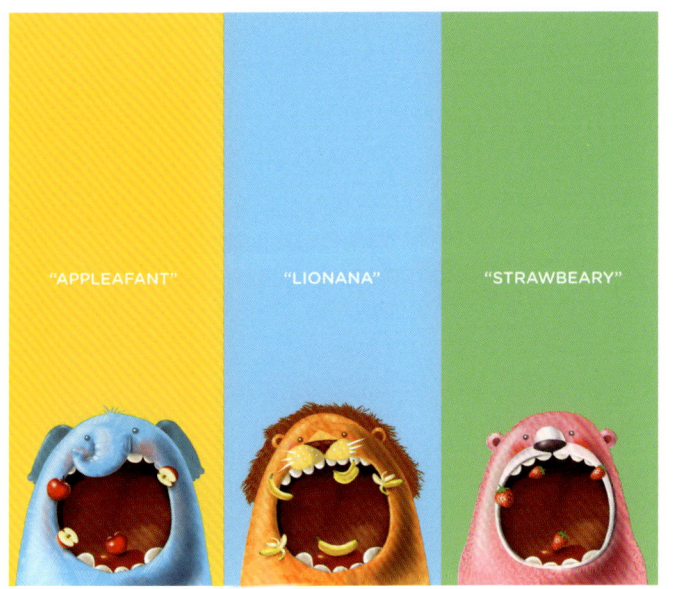

"APPLEAFANT"   "LIONANA"   "STRAWBEARY"

What's Unique?

The 1984 developed the packaging design, including the illustrations. The developed characters based on the flavour – Applelephant, Lionana, and Strawbeary. These characters are created to attract toddlers as they become more aware of what they like and don't like. On the back of the packaging filled with activities for toddlers so as they enjoying the snack they can play with the box as well.

After the success of its older sibling Yummy Bites, the baby rice rusk specialist is back with its product for toddlers called Yummy Bites 123. The rusk rice is bigger in shape, making it easier to handle by toddlers.

COCO FIORI
"SHARE COLLECTION"

Design agency:
Backbone Branding

Designer:
Stepan Azaryan

Client:
Coco Fiori

Country:
Armenia

What's Unique?

The whole packaging design is divided into 6 parts. You can arrange the bouquet as you wish, to present a peace of beauty to the people you love.

The challenge was to create an exclusive collection for Coco Fiori which is a local leader brand in premium flower market. The idea came exactly from the exclusivity of this product. It is a big bouquet of flower, which you can share into six parts and present a peace of beauty to the people you love. Share beauty with Share Collection became the slogan of promo campaign, which exactly describes the idea of this packaging.

FUTURELIFE®
BREAKFAST MEAL

Designer:
Lhente Strydom

Country:
South Africa

Award:
Overall Winner at IPSA Student Goldpack 2016
The Student Worldstar Awards 2016

What's Unique?

The packaging, aiding as a drinking bowl/cup, is created to collapse and expand for ultimate space efficiency and ease of use, for the consumer who lives life on the go.

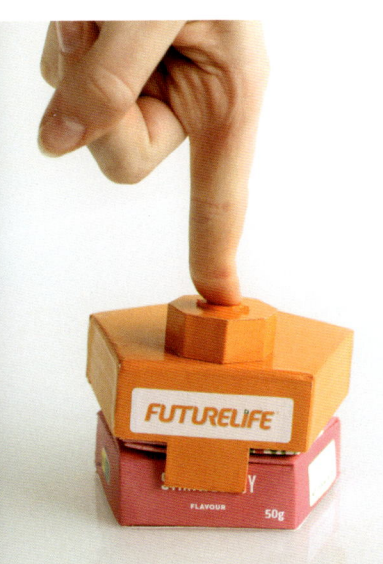

FutureLife® Expand Your Potential is a new and improved on-the-go breakfast packaging design. The concept is based on FutureLife® smartfood's ability to grow the consumer's potential.

MY LITTLE KIKI
PACKAGING

Design agency:
nineteendesign

Graphic designer:
Natalie Poulman

Packaging designer:
Maria Romanidou

Photography:
George Tzortzis

Client:
My little Kiki

Country:
Greece

What's Unique?

The products are "hung" in 2mm solid carton cards and are secured with elastic cords in a variety of colours. Thus, the consumer can experience directly the aesthetic and the quality of the products. The visuals are in the same style, in two colours (white + grey) highlighting their colour range in the best possible way. For the toddler and baby product range, the packaging was adjusted in order to embrace the smaller size of the products. It is made from thinner cardboard, in the same shape and design line.

"My Little Kiki" collection was born, when love and dedication to creation met experience. The owner Kiki Ravanis, having 30 years of experience in children's socks and tights, created her own collection with a unique visual presentation. The packaging design makes the products stand out from those of the competition, and creates an identity reinforcing their dynamic presence in the market.

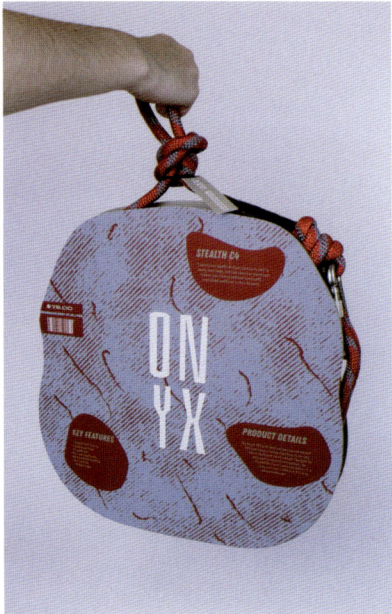

ONYX

Creative director:
Maribeth Kradel-Weitzel

Designer:
Paige Graff

Country:
USA

What's Unique?

The packaging structure has wells for each shoe, which is held in place by rock climbing jugs. This way, the shoes can stand vertically, when held by the rope handle, mimicking the position the shoes would be in when in use. When shoes are taken out, the customer can remove the climbing jugs, to reveal instructions for repurpose. The climbing jugs have holes, so they can be screwed into drywall and used as a small planter, or a catcher for keys and loose change.

In the research, the designer found rock climbers to be very organized with their ropes, shoes, carabiners and climbing packs. She was inspired by the organization of their tools, and designed the structure around how a rope would be wrapped in storage. This interactive shoe packaging, includes a climbing rope that doubles as a hanging feature in stores, as well as a handle for customers after purchase.

CAPTAIN JOHN/
CHEESE

Design agency:
BimBom — ideas for fun

Designer:
Galima Akhmetzyanova,
Pavla Chuykina

Photography:
Pavel Gubin

Country:
Russia

What's Unique?

To make the action easier and quicker a knife can be designed into a package. A tiny turn of the propeller and the cheese will be cut within thin slices.

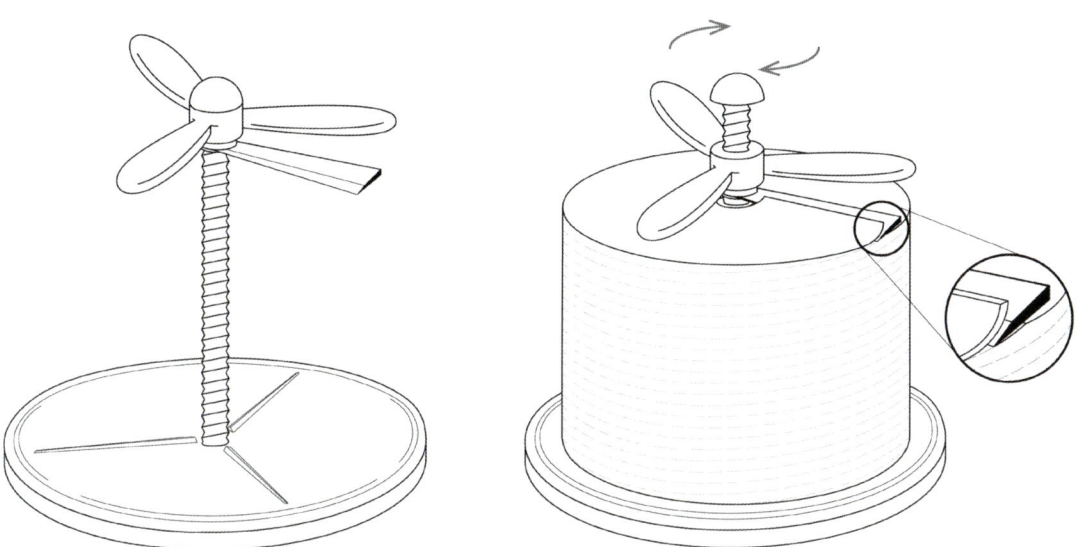

Never underestimate a cheese cutter whose job is to hand cut a wheel of cheese into pieces that are the same size and weight. There is huge responsibility.

CHEERS/MILKSHAKE

Designer:
Pavla Chuykina

Photography:
Yegor Kumachev

Country:
Russia

What's Unique?

Turn the cover and the flavour will pour out. Shake and it is ready to use.

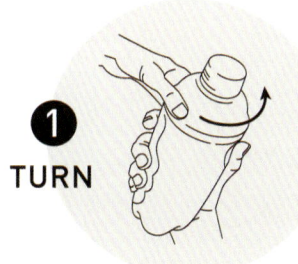

1 TURN

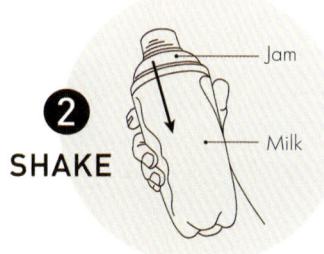

2 SHAKE

3 DRINK

Over consumption of milkshake makes your health stronger. Cheer up yourself with low fat cow's milk drink. Good health! Cheers!

BUTTER! BETTER!

Designer:
Yeongkeun Jeong

Country:
Korea

What's Unique?

The designer replaced its ordinary container lid with a wooden, knife shaped one. This way butter can be easily and quickly spread. This container is not only easy and fast to use but also it makes daily routine of spreading butter more fun and exciting.

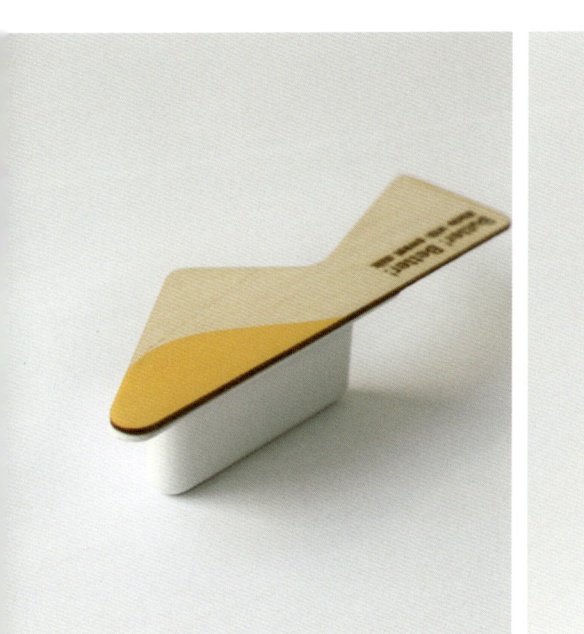

Whenever we eat bread, at the picnic, in the cafe or airplane, we usually use disposable butter. Butter has 4 flavours which allows the user to make a choice, just as he would chose his favourite ice-cream.

ABEEJA

Design agency:
Andrés Guerrero

Photography:
La Industrial

Client:
Salvador Zapata

Country:
Spain

What's Unique?

Through a simple die cut on the label, the packaging can display its wings, showing the iconic image of a bee, a great way of taking advantage of the reduced resources of the client and allows its product to stand above the others. Flying, of course.

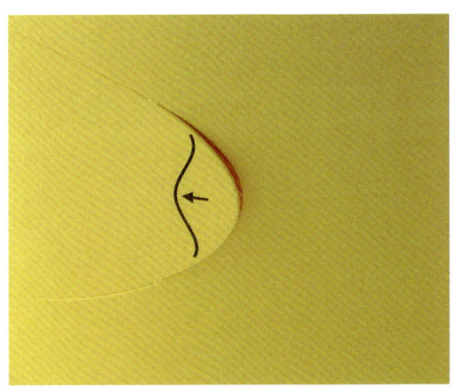

Abeeja is the brand of a local producer who wants to take their honey as far as possible. The idea behind the packaging, to take the product to the limit, begins with the name of the brand that seeks to break borders by linking the words 'bee' and 'abeja'.

HOLIDAY CRACKERS

Design agency:
Goods & Services Branding

Creative director:
Carey George, Sue McCluskey

Designer:
Sarah Rafter, Kristine Planche

Country:
Canada

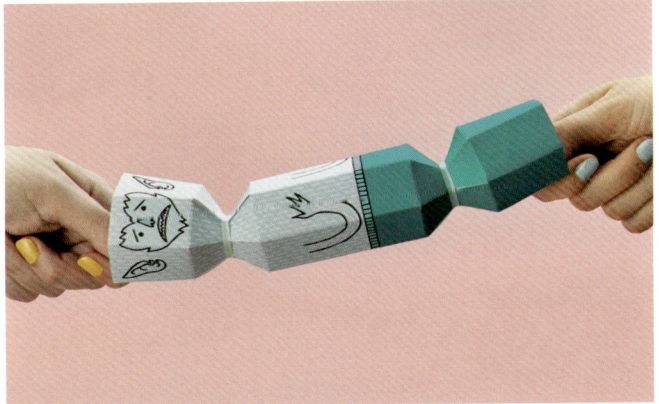

What's Unique?

To complete the experience, the creative team created a web "click" game app featuring the characters, with an addictive high-score notification and results messages such as, "you've disappointed everybody." Gift components included paper crowns and button prizes.

A custom hexagonal box was created to display a holiday story behind each cracker character. The Goods & Services creative team decided to have some fun with holiday crackers. Each cracker includes a story unique to its character: Spain's Tío De Nadal, a friendly log; Iceland's Yule Cat, a notorious feline; Kallikantzaroi, a southeastern European goblin; and Japan's Namahage, a New Year's ogre.

COMON SAVA/
WINE

Design agency:
StudioIN Packaging Ideas

Creative director:
Arthur Schreiber

Designer:
Pavla Chuykina

Photography:
Maxim Kadashov

Country:
Russia

Award:
Red Dot Award 2014 Winner

What's Unique?

It comes from French greeting "Comment ça va" which sounds very similar to Russian word "sova" (owl) and altogether with "comon" it sounds like "come on sova" (which literally means come on owl!). These phonetic synonyms are perfect match because the main hero of this pack is owl by itself.

The trick is that normally tetra-pack wines are standing on lower store shelves and consumers have to look down to see it. And it is getting quite difficult to imagine these wonderful Italian landscapes and juicy grapes in your mind which are pictured on packs when you are looking at it from up to down!

GAWATT EMOTIONS

Design agency:
Backbone Branding

Technical designer:
Karen Gevorgyan

Illustrator:
Narine Manvelyan

Client:
Artur Danielyan

Country:
Armenia

Award:
Bronze Pentaward 2015

What's Unique?

The designers created four various personalities in style of coffee shop without going far from the insight of the brand to give customers a great bunch of positive emotions with just a cup with smile.

After the successful launch of Gawatt restaurant brand, the client decided not to rest on his laurels and promote his brand by using promotional packaging. In this particular case, the designers chose disposable cups as a means of promoting the brand.

CHAPTER 2 • 059

SHAKE MY HEAD - MILKSHAKE PACKAGING DESIGN

Design agency:
Art of Mind

Designer:
Rustam Usmanov

Client:
fakefestival 2016

Country:
Russia

What's Unique?

This really fun and interesting milkshake cup consists of different layers of materials - paper and plastic. The ears of the different characters are perforated for the drinker to tear them out for the extra fun!

The main feature of the design is based on milkshake cup construction and faces depicted on cups. There are 9 different characters and each character has a small family of extra characters for different sizes of cups. Almost every character has perforated ears.

SCRATCH BOTTLE

Design agency:
BBDO

Creative director:
Daniel Schweinzer, Lukas Liske

Designer:
Nicolas Molès

Photography:
Michail Paderin, Christian H. Hasselbusch

Illustration:
Jessica Witt, Marianne Nicolas, Simon Stehle, Lena Dirscherl

Country:
Germany

What's Unique?

It's a party phenomenon: people like to scratch shapes into the characteristic aluminum neck of BECK'S bottles. Based on this insight the designers developed an interactive and playful packaging that turned the whole bottle into an aluminum canvas.

BECK'S is a traditional beer brewery from North Germany. It's served in almost every bar and nightclub throughout the country and has a history in collaborating with artists and musicians. More and more young beer drinkers tend to prefer small breweries and craft beers instead of big players like BECK'S. The designers wanted to find a way to make BECK'S more appealing again to a younger and frequently outgoing audience.

Chapter 3
Establishment of "Interactive Relationship" in Interactive Packaging Design

3.1.1

The establishment of interactive relationship is achieved through experience design in packaging. Themed in "interactive experience", the designer should capture inspirations in terms of product and users. He/She may refer to the product's features and uses, users' behaviour and situation as the starting point and creates resonating and unique interactive packaging design. Common establishment methods of interactive relationship include sensory stimulation, opening way, fusion of feeling and situation, etc.

I. Sensory Stimulation

Aristotle proposed five human senses: visual sense, tactile sense, auditory sense, smell sense and taste sense. The designer may start from these sensory stimulation to convey product information to users. For example, food with fruit fragrance uses smell stimulation to enable users to distinguish the flavours. Wine bottle with braille patterns uses tactile stimulation to provide product information.

See Picture 3.1.1 Work by Alexandra Loginevskaya. This is the packaging design for an anti-hangover remedy. Hangover also brings headache, dizziness and vision distortion. The designer used colour, image and distortion level to show three hangover remedy types. Users could select suitable type through the images on the packaging, which is direct and simple.

See Picture 3.1.2 Work by Eduardo Parás and Constanza Justiniano. The designers created multiple facets on the structure to achieve unique tactile sense for the packaging. Through this "tactile idea", they developed a visually surprising and operation stimulating design plan to activate users' interaction on behaviour and emotion.

3.1.2

3.2.1

II. Opening Way

Packaging is the coat of a commodity, opening the package is the next step after the user get sensory impression. The opening way of interactive design can be designed in 3 stages: before the opening, in the opening and after the opening. Before the opening, the designer should ensure that the opening device is beautiful, easy to use and eye-catching, so that users could find it clearly and open the packaging easily. In the opening, the whole process should start from users' convenience and show humanisation.

After opening, the design should provide users subsequent thinking and emotional interaction about the product. A good opening way for packaging can reflect the designer's professional level, improve the product's grade and enhance the product's competitiveness.

See Picture 3.2.1 Work by Shenzhen Tigerpan Packaging Design CO., LTD. With this unique design, oranges will rise when lightly pulled, which is convenient for picking up and displaying oranges.

See Picture 3.2.2, Picture 3.2.3 Work by Depot Creative and Vert Design. The designers looked to nature for inspiration. To enhance the unpacking experience, the carton employs clever internal mechanics allowing the bottle to rise when the box is opened. As seen by the internal "skin" formed by a folded strip and adhered to the top flaps. The interactive design enhance the product's sense of experience, essential for premium products.

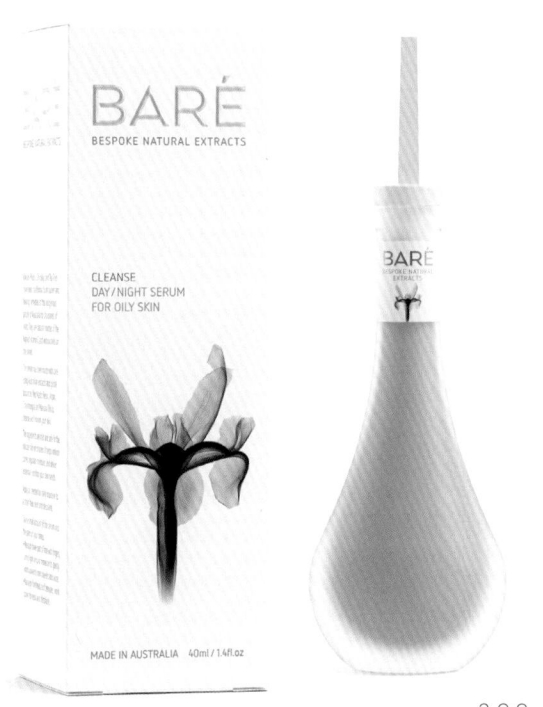

3.2.2

3.2.3

3.3.1

3.3.2

III. Fusion of Feeling and Situation

Today, the requirement to packaging is not limited in use function. Users want emotional satisfaction and enjoyment from packaging. Therefore, interactive packaging design should emphasise feelings and fuse factors such as environment and scenes to endow higher emotional value for packaging, thus achieving dual functions of aesthetics and emotional interaction.

See Picture 3.3.1, Picture 3.3.2 Work by Backbone Branding. Against this packaging for nuts and dried fruits, the designers were inspired by animal's behaviour to store nuts in tree hollows. The packaging was designed as tree hollow. Supreme illustration skills make the packaging look like a real tree trunk. The tree hollow is treated with transparent plastic so consumers could see the nuts inside. Isn't this packaging interesting?

See Picture 3.3.3 Work by Prompt Design. People will always wonder how the product looks like in application environment and what's its effect when choosing products, which is the origin of this design. The designers used a blister pack and put the door handle on the left of the package. The graphics were designed to resemble a real wall texture. With different real wall textures on the package, it attracts the targets and creates huge impact on shelf. The design has won various international design awards.

3.3.3

NONEXISTENT/ VODKA

Design agency:
BimBom — ideas for fun

Designer:
Galima Akhmetzyanova,
Pavla Chuykina

Photography:
Pavel Gubin

Country:
Russia

What's Unique?

Whatever it is, an opticalillusion or a simple camouflage now you see it!

First sight can often mislead and a simple object is more sophisticated than it appears.

Let's have another look and non existent becomes real.

ALCO FREE

Designer:
Alexandra Loginevskaya

Country:
Russia

What's Unique?

The line of "alco free" being presented includes three types of anti-hangover remedy intended for the correction of weak, normal or strong hangover respectively. Indicators of the type are: the colour; the pictogram of the bottle with the estimated amount of alcohol that has been consumed; the image that is distorted stronger, when stronger hangover is. Inside each package of "alco free" there is a set of 10 mirrored bags with the anti-hangover remedy to show you your true face!

A rare holiday, party, birthday or an office party do without alcohol, we drink and laugh, but unfortunately on the morning we have a hangover of different severity levels. The presented "Alco Free" line includes three hangover remedy types to help people to drive away the hangover after a night of drinking.

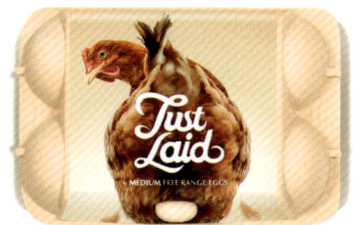
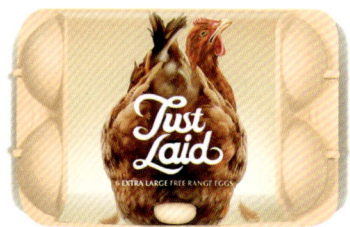

JUST LAID

Design agency:
Springetts Brand Design

Designer:
Kevin Daly

Client:
Noble Foods

Country: UK

What's Unique?

The Just Laid concept uses standard egg box packaging in a unique way that allows the consumers to interact with the pack and design. To further highlight the freshly laid concept, the fastening on the box becomes a representation of the egg itself, allowing consumers to connect with the natural process of the eggs 'being laid'!

The solution was to emphasise the benefit of locally laid eggs to consumers through a cheeky brand name that reinforces the idea of freshly laid eggs straight from the hen. The creative execution is a selection of playful caricatures of hens laying eggs with three different poses/expressions for each of the three variants.

DULUX KIDS SPACE

Design agency:
Springetts Brand Design

Creative director:
Paul Williams

Visualiser:
Stuart Witter

Client:
AkzoNobel

Country:
UK

Award:
Gold Pentaward 2016

What's Unique?

The cans had to be as bold and bright as the colours Dulux creates. Inspired by the visual language of emojis, the designers created a range of happy faces to add emotion to a very functional aisle. Creating a smiling face with the handle not only adds a fun element to the packaging but is also a unique way of making it part of the design.

Kids could personalise the tin with their names and also their colour name "Ducky McDuckface" or "Dinosaur Greeeen" or whatever they wanted to call their room's colour. Or they could simply select from the core range of colours.

Children's rooms are often the most re-decorated rooms in any home, yet all paint cans are designed for adults. The designers created this concept for a range of paints that are targeted at children and adults, to make decorating their rooms fun. The creative concept is Happy Cans - a collection of colours designed for kids.

WILLIAMSON TEA ELEPHANT CADDIES

Design agency:
Springetts Brand Design

Designer:
Sue Bicknell

Client:
Williamson Fine Teas

Country:
UK

Award:
A'Design Platinum

What's Unique?

The caddy would be sold as a gift and so would have to be visually appealing and attractive enough to be retained for reuse (and display) after the contents were used up. Given the elephant is a key part of the Williamson Tea brand, it was vital that the caddy was kept as the shape of an elephant. The outer sleeve had to fit within the brand visual identity and communicate the tea variety and number of teabags whilst showing off the caddy as much as possible.

The designer responded with the idea of producing a range of limited edition elephant caddies. Each caddy would celebrate an event or some aspect of the brand and each have its own character. The contents would remain one of the classic Williamson Tea tea bags.

The designer wanted to develop a range of gift caddies that would take the Williamson Tea brand elephant and transform it into a unique three dimensional icon. The distinctive shape has a strong brand shelf presence and powerfully links the caddy with the brand logo (an elephant). The launch of a range of elephant caddies each with different graphics and character allows the brand to appeal to a wider audience and for the caddies to become collectible. The silver sleeve helps communicates the premium nature of the brand and also allows the caddy to be clearly seen by the consumer.

UT WINE

Design agency:
(Shenzhen) Lingdu Design Co., Ltd.

Creative director:
Liu Zhensong

Designer:
Liu Zhensong

Photography:
Zhou Lijun

Country:
China

What's Unique?

The braille alphabets on the bottle are 3D UV processed to enhance hand feeling and facilitate the visual impaired to know product information, and judge and think about the product. The bottle is designed steady and stable, truly a product packaging for the blind.

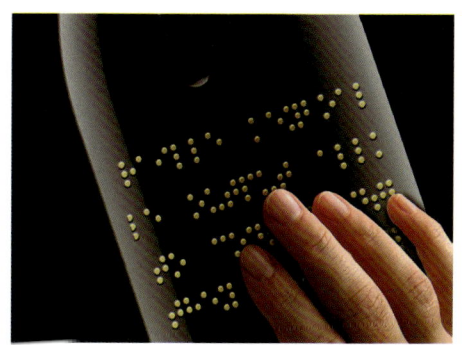

In the beginning of this design, the team made a deep thinking on the demands of blind people. The design aims to show concerns to different groups of people. This design made a deep exploration on the convenience of packaging for this unique group and created this interactive packaging which connects blind people and the product.

HARMONIZATION SHARE

Design agency:
Tigerpan Packaging Design CO., LTD

Designer:
Tiger Pan

Rendering:
Juanjuan Wu

Client:
China Tobacco Hunan Industrial CO., LTD

Country:
China

Award:
2018 Germany Design Award
2017 iF Design Award

What's Unique?

Harmonization Share's idea came from ceiling paintings, or frescoes, an ancient design spanning the East and West. The circle also represents the Chinese philosophers' perfect element of "togetherness." The designer used simple ways of display. Excluding the must-have colour black, he hardly used any ink. This presents a stark but precise and detailed aesthetic standard. This superb cigarillo packaging is notable for its clear and minimalist design aesthetic, which is heightened by its pure white surfaces. Likewise, what is impressive is the manner in which the boxes are unrolled. A fantastic solution.

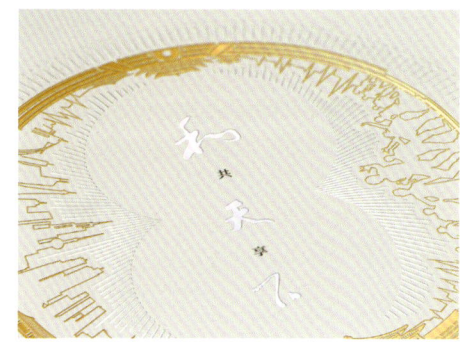

Besides the technique of gold stamping, the package goes through no other printing process while it reaches to a nearly rigorous aesthetic standard. The minimalist way to present profiles of ancient buildings from east to west in a circle, which is exactly a geometric element of Chinese philosophy of harmonization. Find one's true self and you will be enlightened. The simplified presentation method, adopted out of respect for the product itself, conveys the artisan spirit that stands aloof from the world.

CHU'S ORANGE

Design agency:
Tigerpan Packaging Design CO., LTD

Creative director:
Tiger Pan

Rendering:
Juanjuan Wu

Country:
China

Award:
2018 Germany Design Award
2017 iF Design Award
2017 International Design Excellence Awards
2016 Red Dot Award Best of the Best

What's Unique?

With this unique design, oranges will rise when lightly pulled. This assists in picking up and displaying oranges, and symbolises Mr Chu's ups and downs. This design superbly and strikingly revisits the classic aesthetic features of oranges, combining them with sophisticated packaging into a highly noticeable presentation.

Chu Shijian is a legendary, inspirational figure. He started a new business at the age of 74 and became the "Orange King of China" at 88 with incredibly large sales. The designer used wood carvings to represent the respected old man.

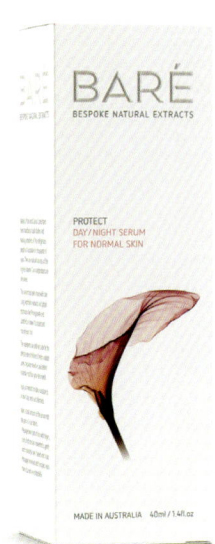

BARÉ ALCHEMY

Design agency:
Depot Creative, Vert Design

Creative director:
Angela Spindler

Photography:
Steven N Meyers

Country:
Australia

Award:
German Design Award Winner 2018
Creativity International Award 2017 Gold
A'Design Award 2016 Gold
Silver Pentaward 2016
The Dieline Awards 2016 2nd Place Winner

What's Unique?

The designers created an own-able asset with the iconic water droplet inspired bottle form and a new delivery system in the form of a leaf shaped dropper. The ridges on the dropper allow for the required amount of serum to be captured and dispensed per application.

To enhance the unpacking experience, the carton employs clever internal mechanics allowing the bottle to rise when the box is opened. As seen by the internal "skin" formed by a folded strip and adhered to the top flaps.

With one of the brands guiding philosophies being biomimicry, the designers looked to nature for both inspiration and direction. The packaging design solution reflects natures inner workings in a very transparent and vivid way, captured through material usage, the frosted glass bottle, the leaf shaped dropper and the brilliant radiographic images. The absolute simplicity and clarity of the packaging design drive its sophistication and harness a well-crafted and well-considered sense of luxury, vital in this segment of the skin care category.

BZZZ HONEY

Design agency:
Backbone Branding

Country:
Armenia

What's Unique?

The designers wrapped the entire jar in paper and came up with hexagon-shaped gold-covered cupboard containers. Placed next to each other, the boxes make an impression of honeycombs. To open the jar, one should first of all unwrap the perforated paper and read the honey production story on the reverse side.

A local honey manufacturer wants to visually depict the multifloral (more than 600 flowers) origin of his honey, which happened to be the point of difference for the product.

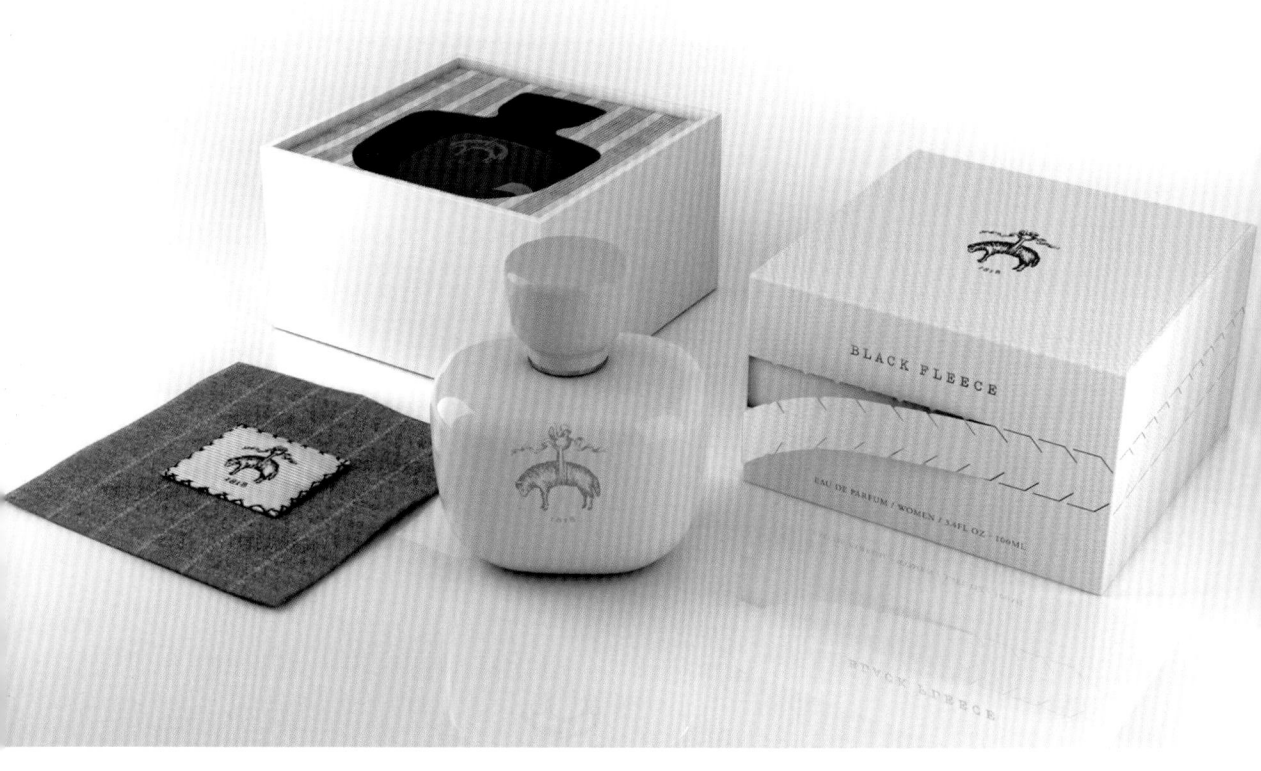

BLACK FLEECE FRAGRANCE PACKAGING FOR THOM BROWNE

Design agency:
Established

Client:
Brooks Brothers

Country:
USA

What's Unique?

A zip-strip on the secondary packaging tears off to remove all graphics, revealing a pure white inner box, containing layers of cut-out felt used in suit-making to reflect the heritage of the brand in tailoring. The packaging is 100% recyclable.

This fragrance packaging for Brooks Brothers sought to combine the traditional world of fine tailoring with the distinctly forward thinking attitude of Thom Browne's black fleece collection. Ceramics and references to office stationary of the 1940s were reinterpreted in a modern way.

THE FAMILY BEEZ

Design agency: Mousegraphics

Country: Greece

What's Unique?

Due to the short lives bees live, The Family Beez needed their packaging to express the importance and pureness of the honey. Each jar is labeled and put into a wooden box. Each box is a limited edition, The Family Beez has produced just 1,350 of each.

The Family Beez is an organic infused honey with a limited production run. Mousegraphics' approach to The Family Beez's packaging is to embrace the authenticity of how the honey is produced.

FOODS CROSS

Design agency:
Mousegraphics

Country:
Greece

What's Unique?

A rare, natural product of an eco-conscious process, meant to be offered in numbered and signed vases. The designers developed the logo design as a careful pairing of cross-shaped lettering (brand name) and the image of a bee, designed by the internationally known illustrator Si Scott. The elongated glass vessel is covered on its upper part and toped by the brand identity elements and relevant information, in a way that allows the synthesis specifics and the collector's data (number) to be clearly visible even when the top is removed. Black, white and red dominate the packaging design in an alternative reference to a pharmaceutical/cosmetics language.

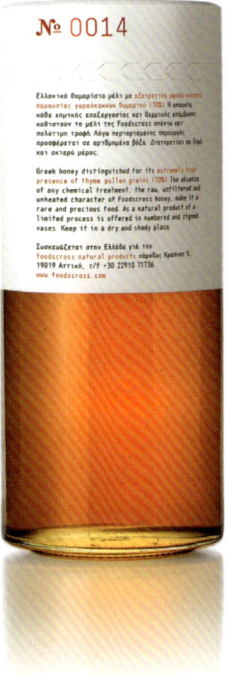

The market for honey is a rising one and as such already filled with a variety of packaging designs. The designers needed to realize a brand identity able to convey this product's specific advantage: it's pure synthesis. This is Greek honey distinguished for its extremely high concentration of thyme pollen grains (80%).

PCHAK

Design agency:
Backbone Branding

Designer:
Stepan Azaryan

Illustrator:
Yenok Sargsyan

Photography:
Backbone Branding

Country:
Armenia

Award:
The Dieline Awards 2018 2nd Place
Gold Pentaward 2017

What's Unique?

Pchak is one of those projects where the idea just stroke the designers and it had to be implemented in reality. While creating Pchak they didn't want to go far from the original idea by including additional design elements. They used the same approach for naming "Pchak" which is the Armenian equivalent of "tree hollow". Pure nature now found its place in our urban daily life.

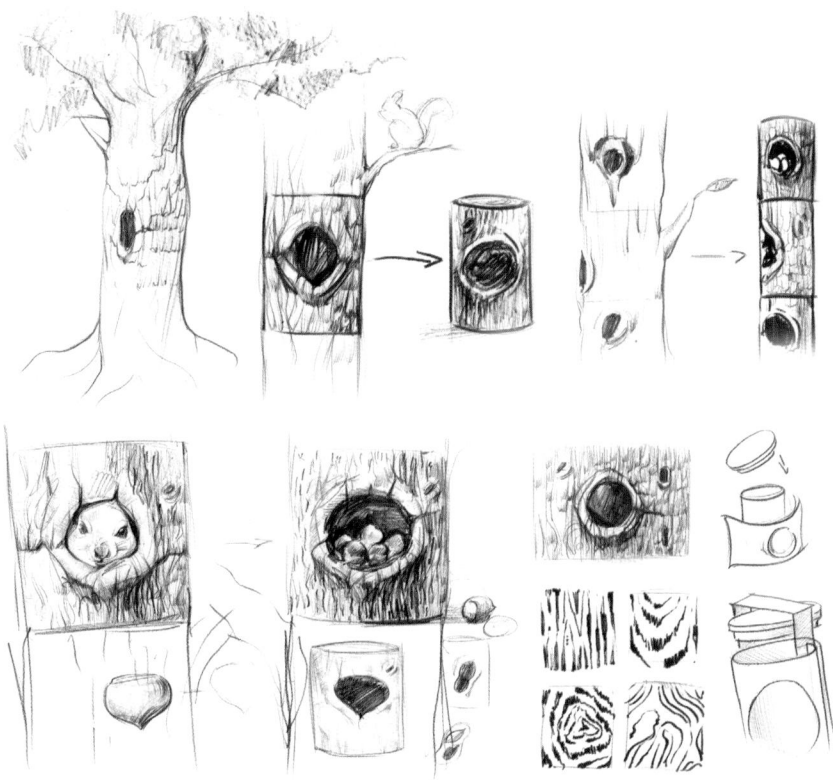

With Pchak's packaging you can enjoy the harvested nut supplies and dried fruits not only in winter but all year long.

SIYA

Design agency:
Backbone Branding

Designer:
Stepan Azaryan

Photography:
Hayk Manvelyan

Client:
SIS Natural

Country:
Armenia

What's Unique?

The trick lies in the real fruit put on the glass which transforms into an entire bottle of juice. The idea is completed with expressive design. The transparent label shows the naturality and freshness of the juice.

What do you see watching at the Siya bottle? Exactly! It is a fruit on a juice glass. Exploring the essential of Siya juice brand, the designer gave it this simple solution – the natural fruit is right on your glass.

BARBERIAN

Design agency:
Opus B Brand Design

Creative director:
Paweł Frej

Designer:
Zuzanna Sadlik, Arkadiusz Tkacz

Country:
Poland

Award:
Red Dot Concept 2017 Honourable Mention
Gold Award Creativity International Award 2017

What's Unique?

Barberian is brand concept for hair accessories that celebrate manhood. The product design itself as well as the choice of colour (just black) are perfectly suited for men with character and inner strength. Barberian's packaging draws attention and tells the brand story despite its small format.

Long facial hair, manbuns or male ponytails have become a symbol of masculinity outside of hipster and metal music subcultures. White collar professionals, entrepreneurs, artists and many more are proud of their hairstyle. Yet there are very few accessories specifically designed for male long hair and men in general. In most cases men have to use female or unisex hair products, which do not embrace their masculinity and virility. Men have inspired the designers to create the brand on the market which is dominated by women's hair care accessories. In order to strengthen the sense of uniqueness, atavistic attractiveness and strength in our men, the designers refer to the desire of any man to be a savage in an orderly, smooth and predictable world of etiquette.

FLORA

Design agency:
Igloo Creativo

Creative director:
Eric Novo García

Designer:
Paola Coiduras Piedrafita

Photography:
Xavier d'Arquer

Country:
Spain

What's Unique?

For this reason the designer decided to create a unique packaging with form of tortoise in which his shell was used as container across which they can see the logo (XXX) created for this special edition. The packaging was 100% personalised for this event.

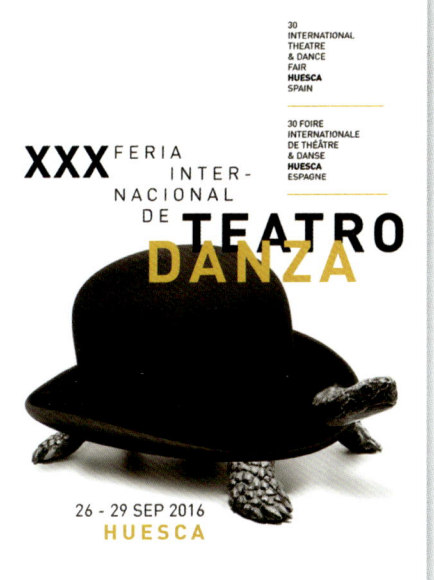

Flora is a packaging design made for the 30 International Theatre & Dance Fair of Huesca (Spain) because of the sale of the commemorative T-shirts for the 30 anniversary of the fair. After designing the image of the event in which a tortoise (Flora) is the protagonist, the designer decided to create a special packaging for the sale of T-shirts doing a "wink" to the design of the poster.

LOS PLAYEROS

Design agency:
Igloo Creativo

Creative director:
Eric Novo García

Designer:
Paola Coiduras Piedrafita

Country:
Spain

What's Unique?

In this collection the designers rescued a character to perform a limited edition of T-shirts. To deliver T-shirts they made shaped ice cream packaging to make more cool still gift and surprise they customers with a fun support.

"Los playeros" is a personal project that the designers have made to welcome summer and thank their customers their fidelity last year. For this project they made a nice illustrated collection of the typical characters that they find in the Spanish beaches in summer.

UNI-PRESIDENT MILK TEA

Design agency:
Mousegraphics

Client:
Uni-President China

Country:
Greece

What's Unique?

The designers developed the most straightforward iconography for each of the product components: a cow for "milk", a tea bag for "tea". When merged these two images produced the most effective, universally understood blend. The photorealistic rendering of the hanging small bag dominates the packaging surface as if such a bag is actually and naturally immersed in the milk tea bottle. Colours are kept equally natural and the coconut flavour product bottle is distinguished by a light and cool shade of green on its bottle.

The client wanted to launch a new brand of bottled milk tea. High aesthetics for a consumer friendly product. The brief the designers were given was rather reserved and straightforward. They opted for the same design language.

BIC

Design agency:
Mousegraphics

Country:
Greece

Award:
Platinum Pentaward 2016

What's Unique?

Each sock packaging is the image of a type of shoe (women's, sports, men's, casual) paired with the appropriate sock. Bic products are differentiated but a larger, and easier to "read" category (shoes) is now providing the unifying principle. Everyday customer choice is transferred to the shoe whereas socks can be of no other brand than the client's.

The client is a well-known company with a strong brand. The designers needed to refresh its image through packaging and also unify several products under one smart idea with functional variations. If socks are dressing the feet and shoes are completing the process the designers took a mental step ahead: they dressed socks in shoes thus creating a playful visual for the most common decision each consumer has to make every day.

ADDA CONTAINER SHOE BOX

Design agency:
Prompt Design

Creative director:
Somchana Kangwarnjit

Client:
Adda Footwear (THAILAND) Co.,Ltd.

Country:
Thailand

Award:
Silver Pentaward 2017

What's Unique?

The ADDA container shoe box design is inspired by the transportation of raw material into the factory and the products to the customers. The design around the shoebox labeled with ADDA name looks robust and is a box-shaped drawer. Moreover it has at 4 corners that easily locked up slots when stacking on top of each other.
By making the shoe box look like a steel container, stacks of shoeboxes can be conveniently arranged as in the container yard. The shape of shoeboxes not only helps in organising at home but also decorative arrangement of the retail display in the shop.

ADDA , the brand of streetwear shoes, has imported a lot of high quality raw material worldwide to produce popular stylish footwear for teenagers. To communicate with modern teen community is a big challenge because of their own society, lifestyle, thoughts and preferences. This leads to a design of ADDA container shoe box.

This newly designed shoebox has successfully impacted the teen community and increased up to four-fold in sales volume.

H4U

Design agency:
Prompt Design

Creative director:
Orawan Jongpisanpattana

Client:
H4U

Country:
Thailand

Award:
German Design Awards 2017 Special Mention
Excellent Communication Design Packaging
Red Dot Communication Design Award Winner
2015 Cat.Packaging

What's Unique?

The designers used a blister pack and put the H4U door handle on the left of the package.

The graphics were designed to resemble a real wall texture such as metal, wood or plastic. With different real wall textures on the package, it attracts the targets and create a huge impact on shelf.

"It is always hard to choose a door handle to match with the wall."

This design concept was developed to solve this problem and make customers decide to buy a door handle easier.

WOOF PACKAGING

Designer:
Maria Romanidou

Photography:
George Tzortzis

Client:
Woof Dog Wear

Country:
Greece

What's Unique?

In order to accomplish the above, a series of hangers was designed with the outline of a dog's head with minimal visual application and brown hue as the basic cutting. The product line includes hangers from thin cardboard for the accessories and solid cardboard 2mm for clothes in 2 sizes.

The company "Woof Dog Wear" was founded in 2011 and its main objective is to design and produce garments and accessories of high quality and aesthetics for pets. Reinforcing this effort, the purpose of the package was to increase promotion and create a complete picture of the brand inside pet shops.

Chapter 4
Design Factors in Interactive Packaging Design

I. Behaviour Factor

Packaging should be people oriented. Behaviour factors in interactive packaging design refers to any behaviour related to the packaging, which is ultimately human behaviour. The primary factor that the designer should consider is to establish interactive relation between users and the product on the basis of possible behaviour which may happen during the use of product.

See Picture 4.1.1 Work by Pesign Design. The inspiration of this wine packaging is Japanese Ninja culture. The labels are designed as outer and inner layers, uncovering the outer layer of the ninja's dress, you will find the beautiful ninja's secret inside, her face turned red, because she drank the wine. The interactive design will provide spiritual pleasure and emotional satisfactions for users, thus promoting the product's sale.

See Picture 4.1.2 Work by Backbone Branding. Since drink will decrease with drinking behaviour, the designer created a simple and dynamic packaging design. Before drinking, red juice and transparent packaging present a full pomegranate. With the decrease of juice, fewer pomegranate seeds are shown until the juice is drunk up. This is a simple yet interesting design.

4.1.1

4.1.2

4.2.1

II. Humanity Factor

Different countries and areas have different local customs and practices. Humanity gives a product deeper artistic value and unique personalised packaging too. For a designer, it is not only a packaging design task, but also the inheritance and output of culture. Only by seeking coherence between packaging design and humanity and by developing deeper spiritual content, will the packaging arouse target users' feelings and gain favours. It is also an important factor to reflect the designer's artistic designing level.

See Picture 4.2.1 Work by Go Peng. ZAOJIAHETAO (Jujube with Walnut) is an emerging popular food in China. This packaging design comes from the homophonic name "ZAOJIAHETAO", which both means Jujube with Walnut and To Marry to He Tao Earlier. The product is personalised as bridegroom and bride to meet the preference of young customers and to be humorous. The packaging has established a foundation for the product's sale and brand communication.

See Picture 4.2.2 Work by Peter Schmidt Group and BBDO. The packaging for the world's first hummus beer symbolises 50 years of friendship between Germany and Israel. Mazelprost is a combination of the Hebrew "Mazel Tov", for "Good luck!", and the German "Prost", meaning "Cheers!". The transparent label emphasises the toast between two countries in a humorous and exciting way. Mazelprost was added to the collection of the Jewish Museum in Berlin.

4.2.2

4.3.1

III. Visual Image

Visual sense plays a dominate role in the human sensory system, so as a way of expression of interactive packaging design, visual image is commonly used. Common visual image elements include typeface, colour, illustration and graphics. The designer could select design elements which may touch users in reference to content's features and convey the product information to users directly and accurately.

See Picture 4.3.1 Work by BULLET Inc. This is the packaging design for a sake brand. The designers were inspired by the famous, ornamental Japanese koi fish, called a living jewel. Red patterns were directly printed on the bottle which resembles the shape of koi. By cutting the box in a koi silhouette, it visually emphasises the image of koi. This vividly designed package will not only stand out in the store, but also be an art piece to decorate your home.

See Picture 4.3.2 Work by Studio Sonda. The labels of this wine collection are designed to speak out about the importance of understanding Nature. Vintage year is always denoted on the bottles, but how many of us really understand its meaning? It actually witnesses the natural conditions in which the wine matured, such as weather, amount of precipitation, etc. Flavours are different due to different natural conditions. In cooperation with the meteorological service, the data on weather conditions in the territory of vineyards were collected, and the amount of precipitation shows graphically how much weather actually affects the diversity of nature every year. Circles show the amount of rainfall in a particular month. The result are labels that give consumers the possibility to easily compare the different vintage years.

4.3.2

4.4.1 4.4.2 4.4.3 4.4.4

IV. Structure Design

Packaging structure design is a three-dimensional design. Besides process, material, cost, transportation, sale factors, the designer also needs to consider users' physical and psychological factors and integrate formal beauty design rule and design requirements to design the container's internal and external structure. The designer could refer to real forms and use comprehensive design of point, line and plane to create impressive structures.

See Picture 4.4.1 to Picture 4.4.4 Work by Backbone Branding. The designers were inspired by bitten apples. The bottle was designed based on biomimicry principles: bottles like bitten apples are put one after another which serves as a visual feature for the shape of the bottle. This way of arranging the bottles allows for a significant save of space both on shelves and in a container. In this design, the designers considered users' psychological activities, user habits and utilised aesthetic principle to extract visual features of packaging design. Finally, a unique interactive packaging is presented.

See Picture 4.4.5 to Picture 4.4.8 Work by Marija Brkic. This chocolate packaging design is inspired by the ancient civilizations of Mayans and Aztecs, present by pyramid and ancient fresco-painting. The designer created ten different shapes to carry chocolates, which assembly to a whole pyramid. This packaging has an educational role: through the game and the consumption of chocolate, kids can learn about different geometric shapes, history, architecture, art and mythology of the ancient civilizations of Central America.

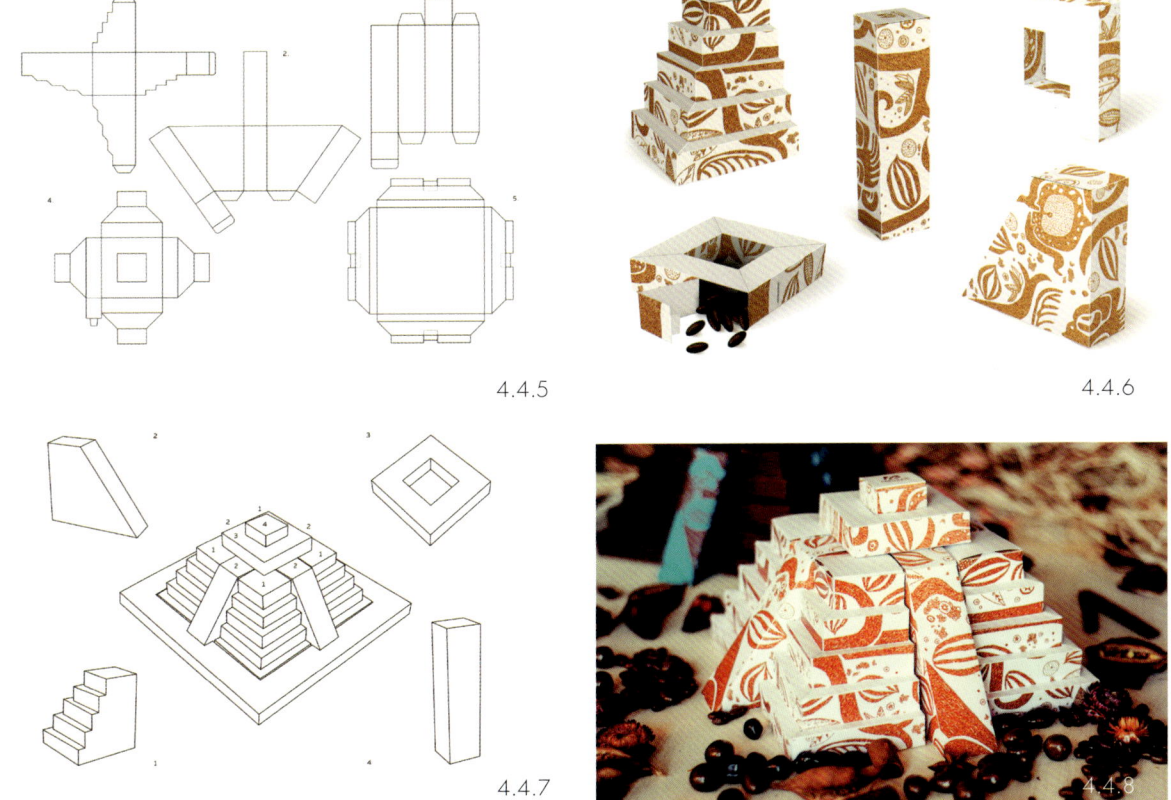

4.4.5

4.4.6

4.4.7

4.4.8

4.5.1

V. Material Selection

Interactive Packaging emerges with the birth and development of new technology and new materials. Materials with different properties will endow the product quite different effects. The designer should seek correct packaging material for the product and achieve interaction with users on the basis of its functions. For example, food packaging made of edible material enables users to eat the packaging when they have finished the food. On one hand, it adds fun for users; on the other hand, it reduces waste pollution from packaging. More and more users are paying attention to their influences on environment. Therefore, selection of natural materials, and reuse, recycle and degradable green materials will improve users' favourability to the product and establish a positive brand image for it.

See Picture 4.5.1 Work by BBDO. Big parts of the world's bee population are endangered. Aimed to protect bees, BBDO came up with an upcycling design idea that turns wood packaging into a bee hotel: beebnb. It works by combining the wooden box with biological straws – creating a small beehive. You could put it to a sunny place, thus providing bees a refuge. All materials used are natural and sustainable, which protect bees as well as the environment.

See Picture 4.5.2 to Picture 4.5.4 Work by Miriam Köni. The target group for this toothpaste is children. To make it more interesting and attractive, the designer drew various, colourful characters. It is particular that the packaging has two layers: inner plastic and special outer paper that dissolves when getting wet. If the tube is put under water, the package band dissolves, where the former package band was placed and cheerful characters will appear.

4.5.2

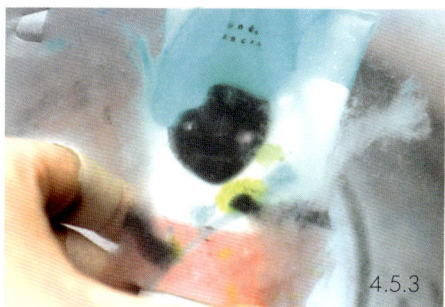

4.5.3

4.5.4

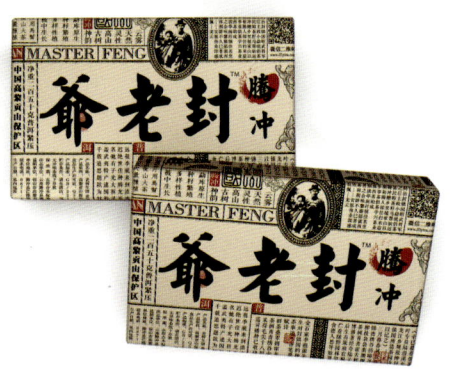

FENG LAO YE

Design agency:
Tigerpan Package Design CO., LTD

Designer:
Tiger Pan

Client:
Yunnan Tengchong Gawagapu Tea Co.,Ltd.

Rendering:
Juanjuan Wu

Country:
China

Award:
Silver Pentaward 2015

What's Unique?

"Who is Feng Lao Ye?" This is the question that the designer want to hear from consumers. Feng Lao Ye was a famous squire in Yunnan in the period of the Republic of China. He and his son were remembered for introducing the large-leaf Pu'er tea to the picturesque Gaoligong Mountain reserve area, which later became a source of rare Pu'er tea from old tea trees. The design adopts the newspaper format in the period of the Republic of China, telling the long stories between Feng Lao Ye and Pu'er tea.

The client came to the designer after seeing some of their previous works. The product of black tea is famous in the local area but not out of it. Mr. Pan named it by Feng Lao Ye before the design is finished. The client likes it much as it memorizes the creator of the tea in Yunnan. Also Lao Ye in Chinese means people who are powerful and aged or in their middle-ages These people are exactly the target consumers of the client.

EVA NEWTON JOBS

Design agency:
Tigerpan Package Design CO., LTD

Designer:
Tiger Pan

Client:
HongheCangying Agricultural Technology Development CO., LTD

Rendering:
Juanjuan Wu

Country:
China

Award:
2018 iF Design Award
2017 Red Dot Award Winner
2017 Pentaward Nomination Award

What's Unique?

It is never doubted that being interesting itself can bring huge value to products. What is "Xia Niu Qiao"? It is actually a combination of the first Chinese characters in names of Eve, Newton and Jobs. The designers try to use the exaggerated and sarcastic expressions in illustrations to present these three people linked to the apple, thereby trigger the fast perception and fun imaginations of the "Xia Niu Qiao" brand among the audience. Just as the advertising slogan states: the fourth apple that changes the world. The design aims to take up a place in the memories of the audience and make it something that people have to make an effort to forget. It is a nice thing.

This is a package design for fresh apples. The inspiration came from three great people who are related to apples, Eva from Eden, Newton with the falling apple, and Jobs with the Apple. The portrait of the three is never easy as the designers try to convey a string of fun by the package, which definitely helps attract attention of consumers and distinguish the brand from its competitors.

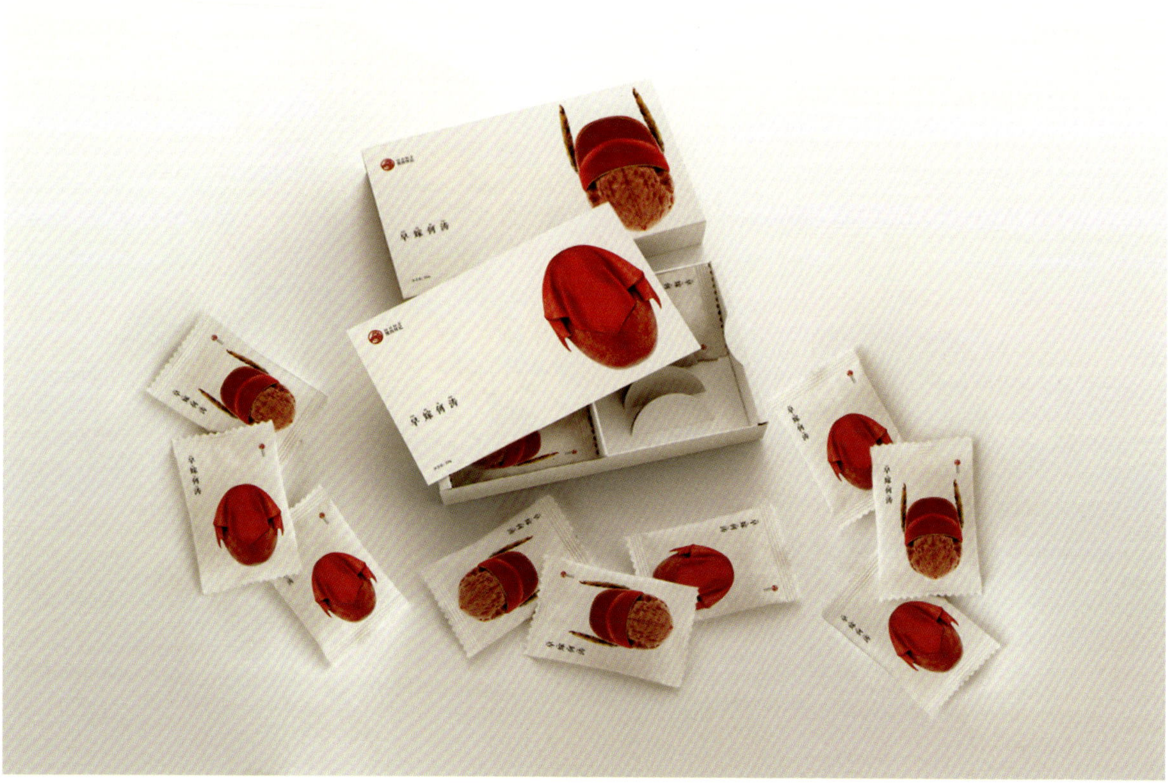

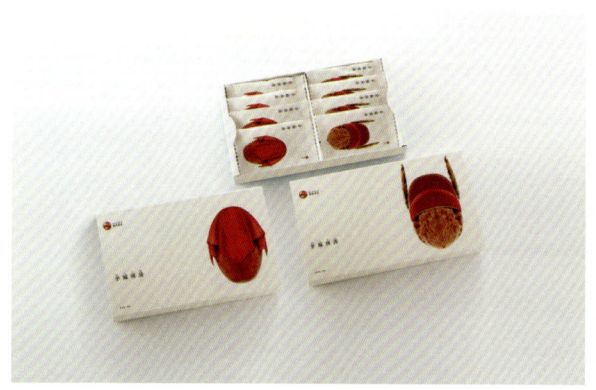

ZAOJIAHETAO

Design agency:
www.xiangaopeng.com

Creative director:
Go Peng

Designer:
Go Peng

Photography:
Xu Yan Liang, Zhang Long Yuan

Country:
China

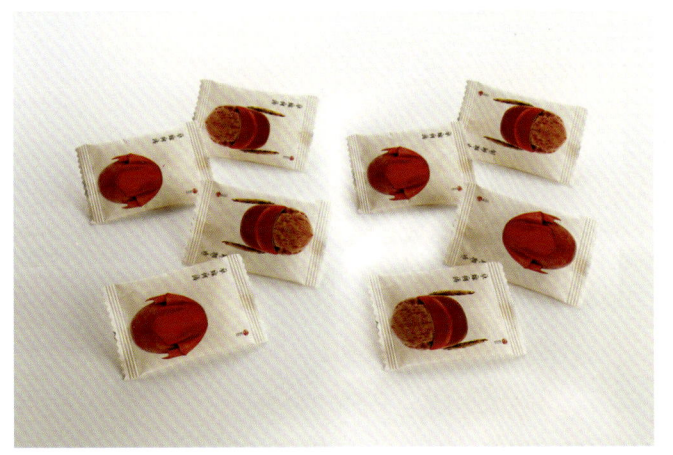

What's Unique?

The packaging personalised the product (jujube and walnut) and used unique homophonic Chinese creatively, combining the product and the images of bridegroom and bride, creating interaction between product and packaging.

ZAOJIAHETAO (Jujube with Walnut) is an emerging popular food in China. This packaging design comes from the homophonic name "ZAOJIAHETAO", which both means Jujube with Walnut and To Marry to He Tao Earlier. The product is personalised as bridegroom and bride to meet the preference of young customers and to be humorous. The packaging has established a foundation for the product's sale and brand communication.

BASKETBALL CRAFT BEER SELECTION

Design agency:
kissmiklos

Manufacturer:
Karl Micro Brewery

Client:
Erzsébet Alehouse

Photography:
Eszter Sarah

Country:
Hungary

What's Unique?

The project resulted in three colourful labels with the players' simplified portraits whose eyes are replaced by their jersey numbers.

This is a special craft beer series in packs which is purchasable only in Erzsébet Alehouse in Kaposvár. The owners of the alehouse would have liked to create some beer labels about their favourite basketball players who played or are currently playing in the basketball team of Kaposvar.

WENSKY FOLCLORE

Design agency:
Labis Design - Artefice Group

Creative director:
Henrique Catenacci

Designer:
Maikel Morais

Photography:
Labis Design

Country:
Brazil

What's Unique?

The label graphic line stands out from the traditional. In this project, Brazilian folklore characters are not only icons nor mascots, they are the own label. The figures were abstracted with the creation of a single trait, which reinforces the characters' personalities.

The goal was to create a special line with strong personality, so that the consumer could associate immediately the type of the beer to the character.

MAZELPROST

Design agency:
Peter Schmidt Group

Creative director:
Dennis Dominguez

Designer:
Juliana Fischer, Anette Haid, Stephanie Rieckmann, Florian Schaake

Client:
Embassy of Israel, Berlin, Germany

Country:
Germany

What's Unique?

The beer's name equally links the two countries together as Mazelprost is a combination of the Hebrew "Mazel Tov", for "Good luck!", and the German "Prost", meaning "Cheers!". Its transparent label, seal and colour scheme reflect current trends within the craft beer segment to give Mazelprost a contemporary image. In December 2015, Mazelprost was added to the collection of the Jewish Museum in Berlin.

The project is a collaboration with BBDO Berlin and BBDO Tel Aviv. The packaging for the world's first hummus beer symbolises 50 years of friendship between Germany and Israel, and it does so in an exciting and humorous way.

NICHOLAS AND KRAMPUS WINES

Design agency:
Goods & Services Branding

Creative director:
Carey George, Sue McCluskey

Designer:
Christian Blau

Country:
Canada

What's Unique?

The scalloped butcher-paper wine labels, featuring stylised depictions of Nicholas and Krampus, are hand-wrapped on the bottle and tied with a decorative string.

The wines arrived at their destinations packaged in an understated black and white box that gave a little more information about the characters and their origins -- and an assurance that the wines would be enjoyed by both the naughty and the nice.

The design presents a naughty/nice dichotomy: one white wine and one red wine (both 100% Ontario – VQA Niagara Peninsula, to be precise), dressed up as Nicholas and Krampus, those standbys of Alpine folklore (Nicholas is the basis of Santa; Krampus is a horned punisher who has become increasingly popular in North America in recent years).

HOLLYWOOD LIMITED EDITION - MAKEUP GIFT SET

Design agency:
www.nx-design.net

Art director:
Guozheng Jiang, Dandan Chen

Designer:
Ke Zheng, Yue Ni

Client:
MAX FACTOR

Country:
China

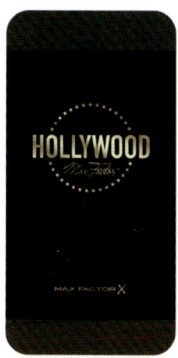

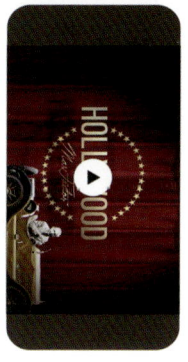

What's Unique?

in honor of the founder, Mr. Max Factor, and Hollywood in the 20s we set the distinctive visual theme, and design the packaging as a retro film box, for the limited edition gift box. Inside the gift box, three makeup products are deduced into a three-act play from an old film featuring Mr Max Factor and the stars whom he put on his makeup. There is a line on each product that explains "beautiful makeup". And we attach tickets to the 20s film and a real film strip on the back of the lid.

At the back of the Hollywood ticket in the box there is a QR code of brand H5 Video to let the product itself bringing the secondary spreading for brand.

Max Factor is a century-old brand of makeup from the Hollywood in 1920s. As time goes by, Max Factor's original gene of Hollywood has faded away with the transfer of brand owners. It has always been a pity that most consumers, especially in China, don't even know that Max Factor, the brand name is the very same name of the founder's. Through the creative design, innovative way of craft production and spreading of the flow to promote the brand image, strengthening the brand assets, pulling the new growth.

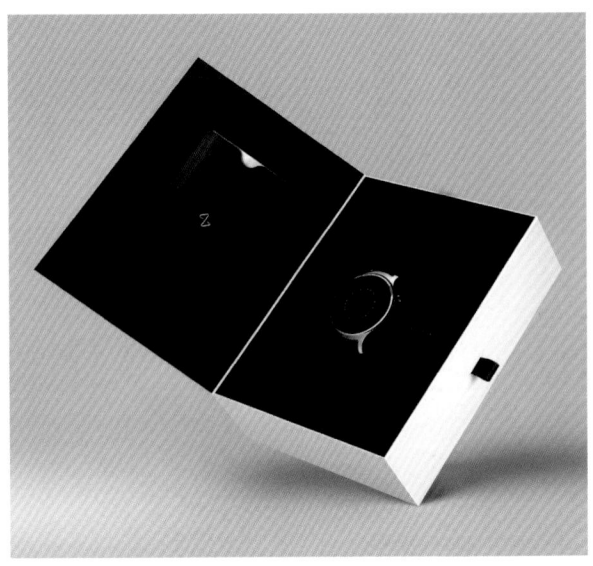
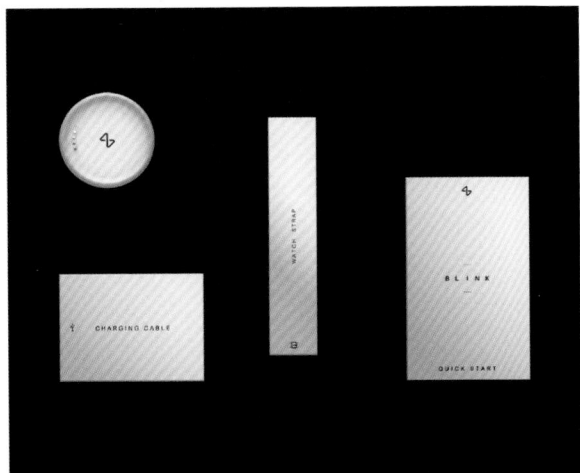

BLINK WATCH PACKAGING

Graphic designers:
Nikhil Bapna, Mani Singh

Industrial designer:
Ishaan Dass

Client:
Blink

Country:
India

Award:
Semifinalist in the Commercial Design Category at Adobe Design Achievement Awards 2017

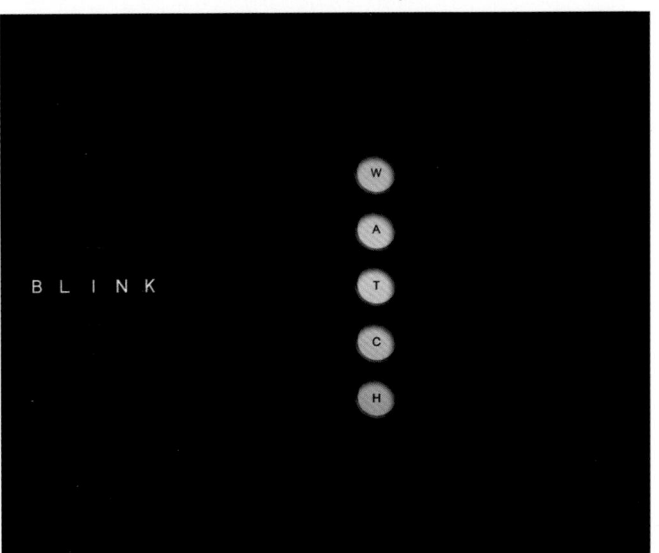

What's Unique?

The unique experience doesn't rely on attractive colours and images, but the minimalist design. The packaging has two layers: the inner layer is white box with black letters; the outer layer is a black sleeve with 5 holes. When the user pull out the box, the changing letters will compose different words.

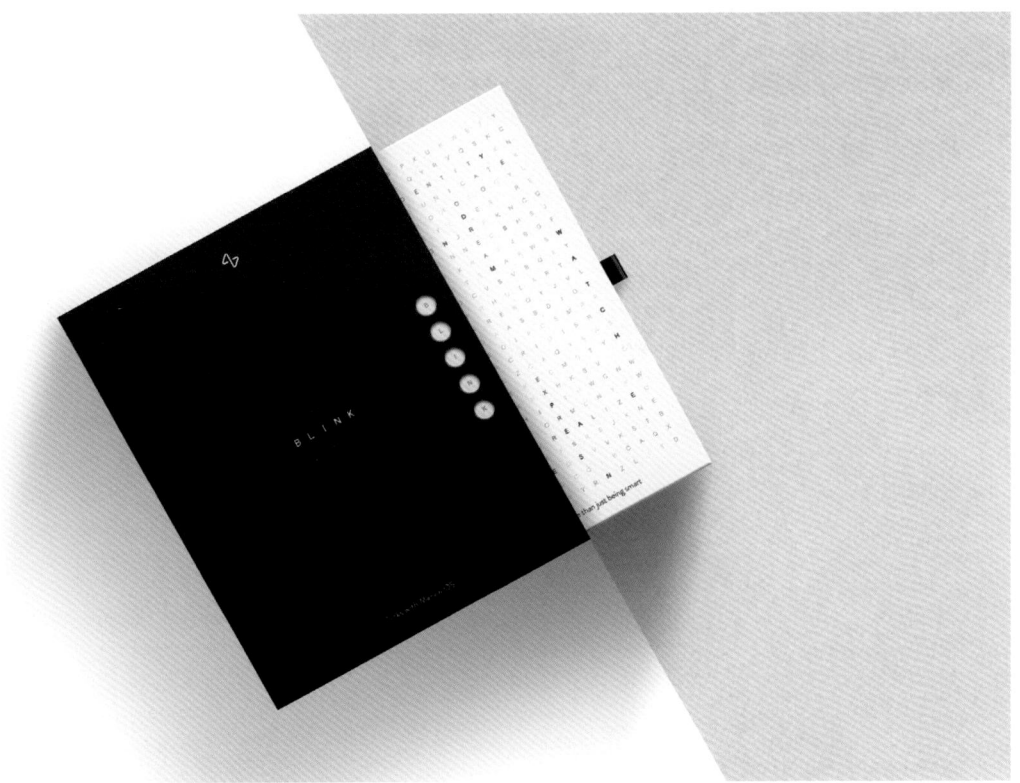

Blink is a Bangalore-based modern-day consumer technology company that makes smart devices in home and lifestyle space. The purpose of this design was to create a simple yet iconic experience that will reflect the brand of 'Blink' and its values. It has to be artistic and visually appealing as the device inside.

REAL BANANA MILK
PACKAGING

Design agency:
Dongwook Yoon

Photography:
Korea

What's Unique?

What if you can drink banana milk like peeling a fresh banana to take a bite? You can wrap off a bottle of banana milk like peeling a real banana. By reflecting the emotion and the memory of peeling a banana, this packing idea can highlight the freshness and sweetness of the banana milk as you take a bit of the real banana.

Peel a Banana and Drink. By expressing the shape of the standing cut-off banana and the image that looks like the real banana surface, it gives the feeling of eating a real banana.

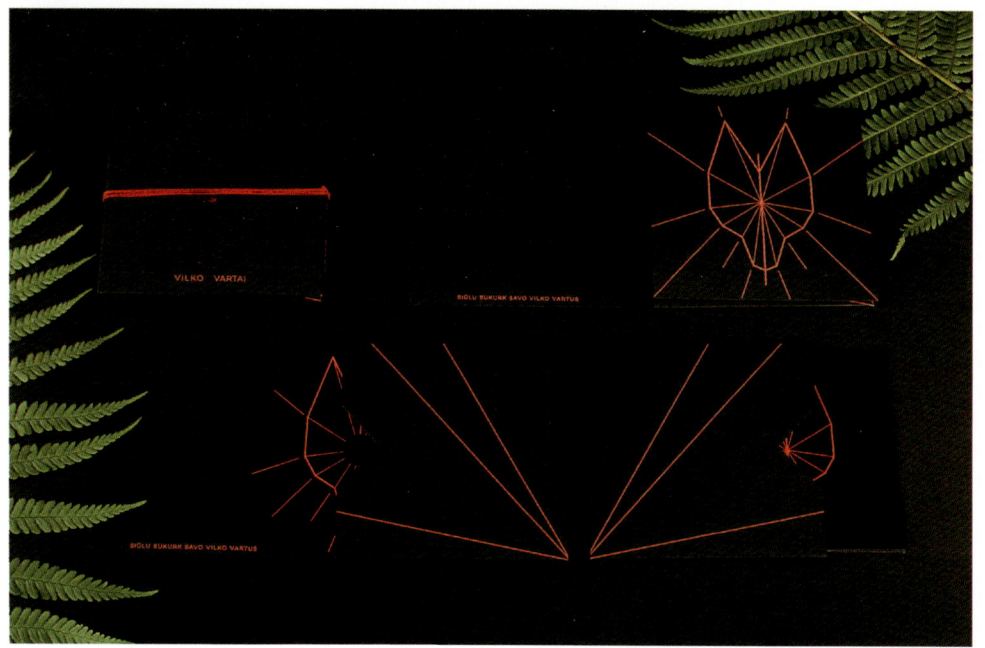

LIMITED EDITION WOLF GATES MUSIC ALBUM

Designer:
Ernesta Vala

Client:
Neo folk band Skylė

Country:
Lithuania

Award:
NAPA award (Best Lithuanian Artisan Package)
Adrenalin Bronze Arrow

What's Unique?

The music album owner can create his own cover with a string through the holes.

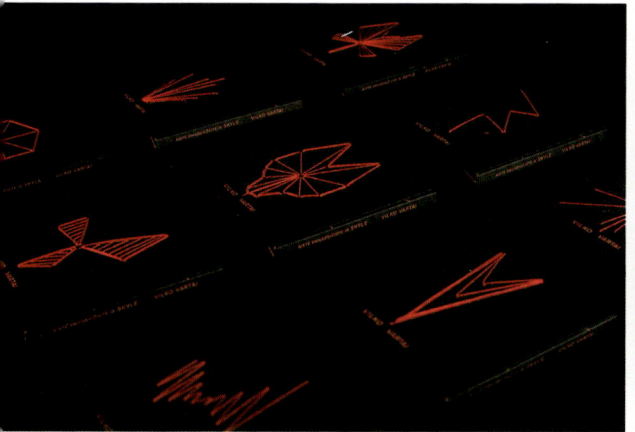

In design process the designers coded a man's life as a line where he is a master to pull the string how he wants. In the end they have two music album covers for mass production and limited edition. In both designs you can create your own visual with added string, while connecting the holes.

CHAPTER 4 • 149

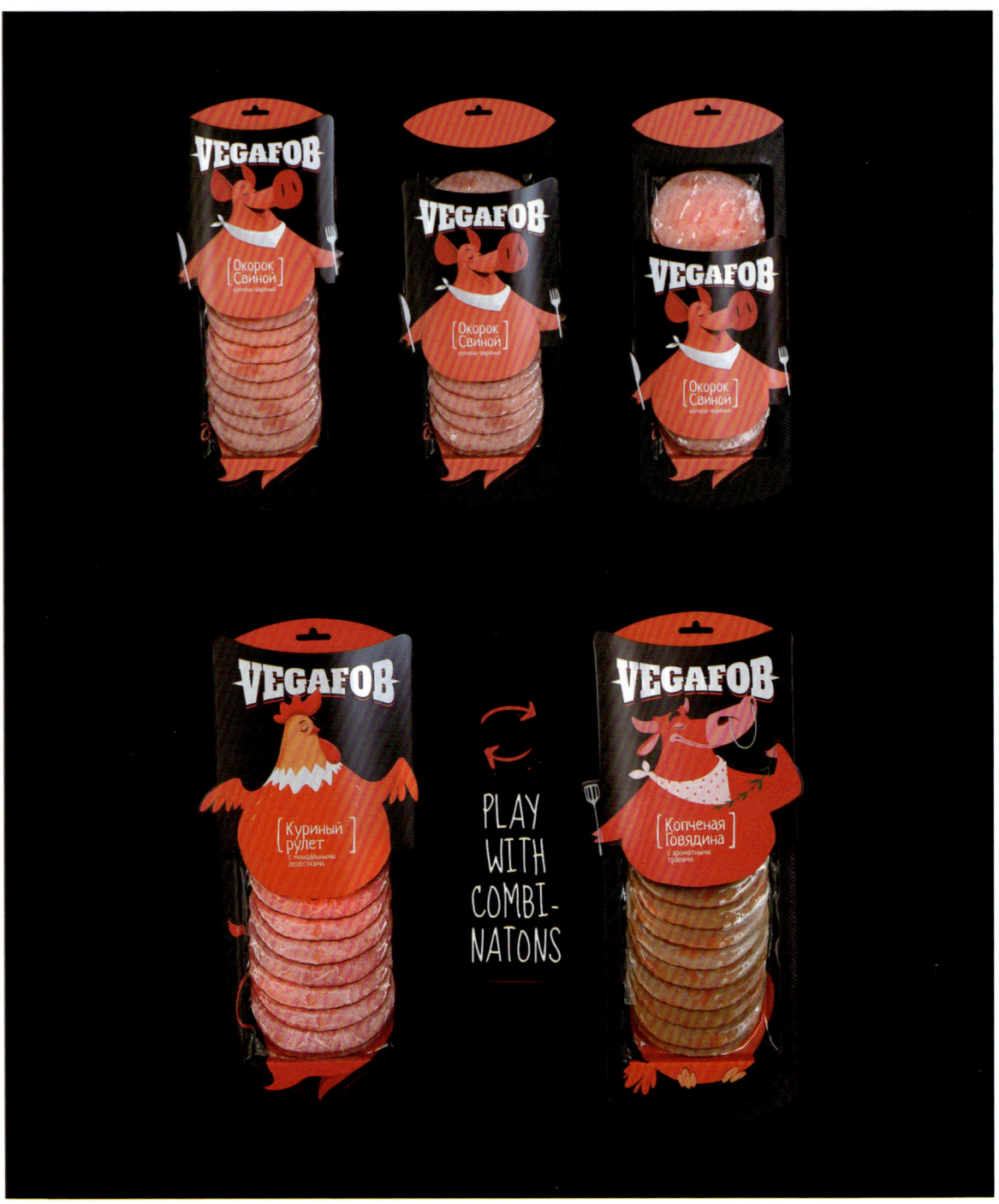

VEGAFOB

Design agency:
Art of mind

Designer:
Rustam Usmanov

Rendering:
Andrew Sylka

Country:
Russia

What's Unique?

The feature of the design in conjunction with the design, as well as an additional printing inside gives an understanding of the idea in the process of interaction of the buyer with the product.

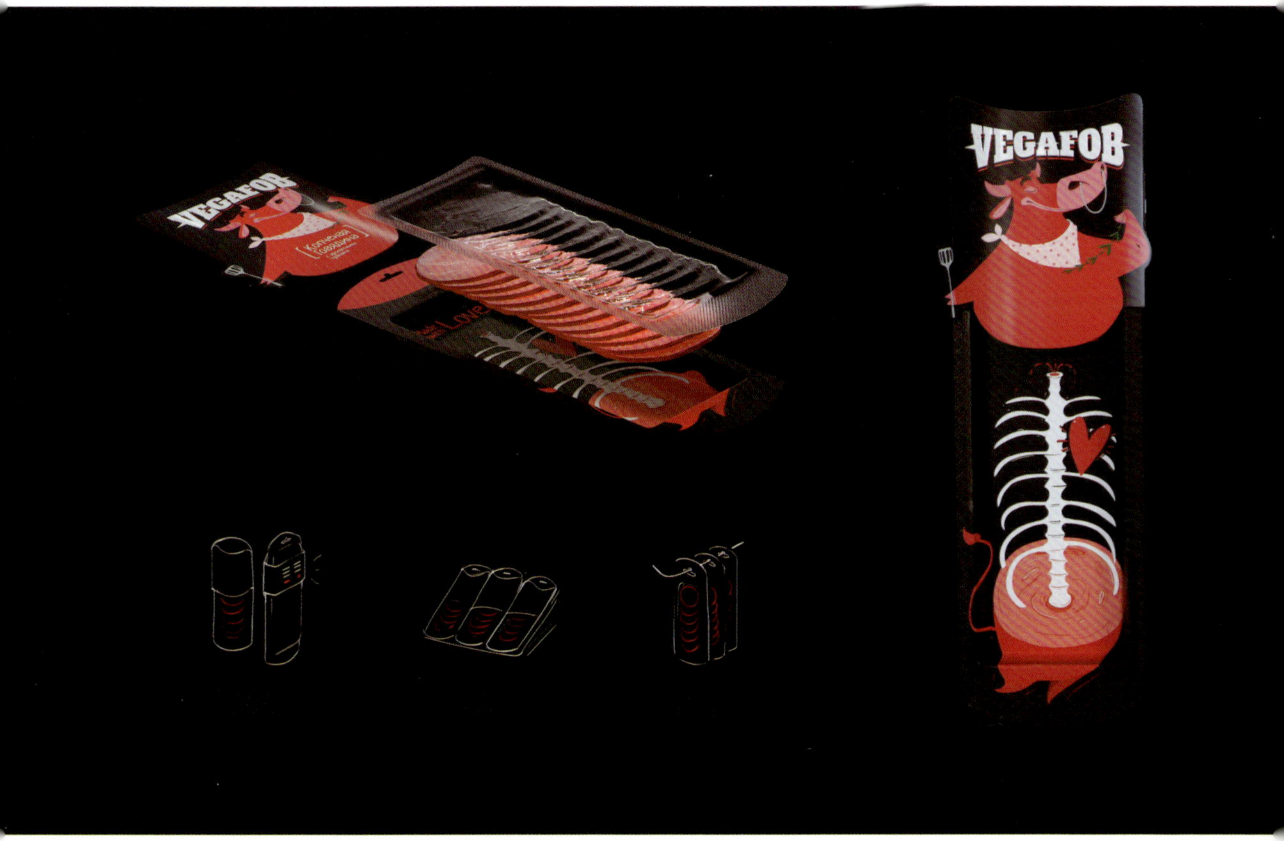

An unusual approach in the design of sausages. For lovers of meat with a sense of humour.

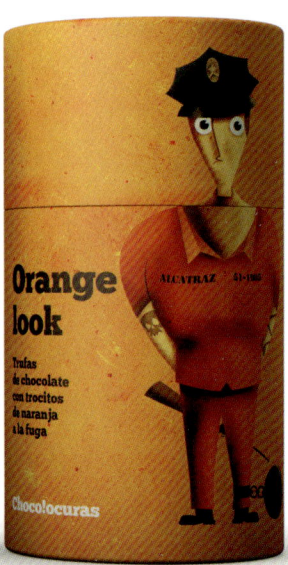

CHOCOLOCURAS

Design agency:
Supperstudio

Client:
Fresh and Good

Country:
Spain

Award:
FAB Awards 2017 (Silver)
Best Award 2017 (Silver)
Premios Salón de Gourmets 2017
Gold Pentaward 2017

What's Unique?

It's a game, humour and fun: Who's who? The thief is the policeman, the lion is the hunter ... A different pack for a very competitive category. A design that changes only if you turn the tap. Two characters exchange their bodies and their heads to create a new and different one. The colour use and the naming of each product give notoriety and presence at the point of sale. Chocolocuras pursues impulse and a different proposal.

The product is selling in leisure stores specialised in books, music and technology. The packaging must surprise and connect with the audience that is looking for entertainment. The container is a cartonboard tube packaging covered with paper.

LA CASITA BLANCA

Designer:
Laura de Miguel, Cristian Varela,
Maria Romero

Country:
Spain

What's Unique?

Against the white packaging, the rich brown chocolates stand out boldly. The name of the chocolates is hidden, and a small tab can be pulled to reveal it.

La Casita Blanca is chocolate with a bit of mystery, elegant and striking. Chocolates and their packaging inspired in "LA CASITA BLANCA", the most famous Meublé in the history of Barcelona, founded in 1912. During the postwar years, the attention to detail turned this place into the perfect shelter for all sorts of clandestine love. Its name had origin in the large linens drying in the sun, that you could see at the roof of the building.

ORANGINA SPRIAL PEEL

Designer:
Yuko Takagi

Photography:
foton

Client:
Suntory

Country:
Japan

What's Unique?

You get a different message with every bottle so collect them all! Please enjoy!

The designer simply transformed Orangina's iconic image "the orange peel" into a fresh-looking bottle. When you peel the label, messages will come out in a spiral.

ZARA KIDS

Design agency:
Lavernia & Cienfuegos

Creative director:
Nacho Lavernia, Alberto Cienfuegos

Client:
Inditex (Zara)

Country:
Spain

What's Unique?

The designers used cardboard tubes to rotate the lid so that the eyes of each character changes with each turn. A nice little touch is that the positioning of the lids is not fixed to the base of the tube, which means that each reference lines up differently at the point of sale.

ZARA KIDS is a fragrance for children between the age of 6 and 9. Speaking to a child's visual language, the designers created two fun and attractive characters with special interest in giving the characters on the packaging a playful spin (literally).

SQUEEZE & FRESH

Design agency:
Backbone Branding

Designer:
Stepan Azaryan

Country:
Armenia

Award:
Pentawards 2016 Concept Silver

What's Unique?

The label of the juice cup is simple but dynamic and it acts with you when you use it. It makes you excited, gives creative mood and makes you feel a part of something.

Interactivity has become a powerful competitive advantage in today's market. Creating the "Squeeze & Fresh" juice cup the designers shared the idea of the package, which communicates.

FM SHOWER GEL

Designer:
Aleksei Pashnin

Country:
Russia

What's Unique?

This package concept was designed to encourage them to reveal their vocal potential. The images of the famous singers on the front of the label will help them to adjust on the right mood and throw away the diffidence. And sponge top part of package will serve as an additional convenience while washing.

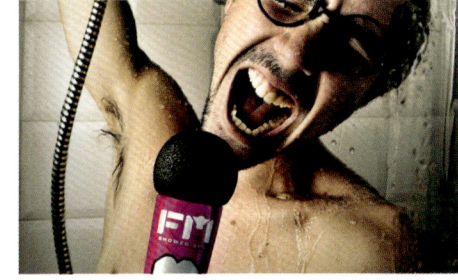

Do you like to sing in shower? If your answer is "yes", this shower gel package is exactly what you need. Throw away diffidence and feel yourself like a superstar! But be careful, because your neighbours might knock on your door not to take your autograph!

ANTISMOKE PACKAGING

Design agency:
Reynolds and Reyner

Creative director:
Artyom Kulik, Alexander Andreyev

Country:
Ukraine

What's Unique?

As part of an anti-smoking campaign, the designers wanted to draw the link between smoking and death through a compelling design. By drawing the cigarette packaging as a coffin, they've created an obvious visual message – smoking kills – in a way that doesn't need supporting text.

Smoking-related illnesses continue to be a leading cause of death around the world. This conceptual packaging illustrates just how close this problem is to all of us, whether or not we're actually smokers ourselves. Each smoker carries death in his or her pocket, wherever they go.

BZZZ PREMIUM HONEY

Design agency:
Backbone Branding

Designer:
Stepan Azaryan

Country:
Armenia

What's Unique?

The designers have created the cover of wood looking like an improvised beehive where the can with honey is hidden. The initial sketches made during the briefing session with the client served as packaging concept for the product. Designed in a notion of biomimicry, the wooden hive left no room for alternative concepts and made the way to the technical execution.

The project lies in developing limited edition of honey jars as if they were a business gift. What container can pretend to enclose honey best? The designers came back to the nature and got inspired through real shape of a hive.

YAN

Design agency:
Backbone Branding

Creative director:
Stepan Azaryan

Illustrator:
Armenak Grigoryan

Realisation:
Lilit Arshakyan

Client:
SIS Natural

Country:
Armenia

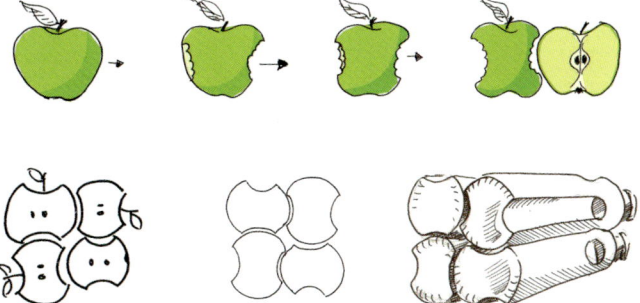

What's Unique?

The concept of Yan bottle was designed based on biomimicry principles: bitten apples put one after another which serves as a visual feature for the shape of the bottle. This way of arranging the bottles allows for a significant save of space both on shelves and in a container.

Label with an application of a stylised calligraphic inscription "Yan", as a thin grass reed reminding the nature. Recycled paper is used for glass bottle label. The brand concept is "Organic in Everything" which has served as an impetus for the new line of Yan brand.

Backbone Branding's collaboration with SIS Natural juice manufacturer started years ago, by developing a new juice brand Yan. As a result, the designers developed a unique glass bottle which immediately put the product into premium category.

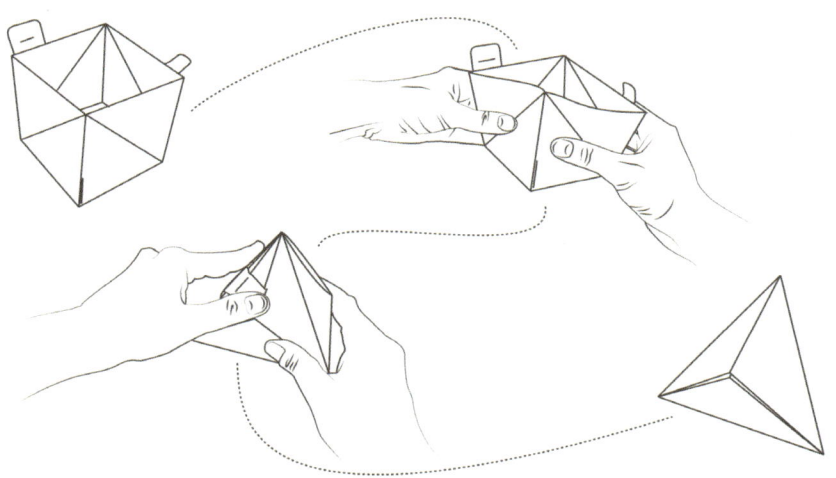

PAPRIKA GROW BOX

Creative director:
Balázs Tóth

Creative director:
Balázs Tóth

Photography:
Balázs Tóth

Award:
WPO World Star Student Award/ Certificate of Recognition

Country:
Hungary

What's Unique?

By dividing the two open sided boxes, one part becomes the pot of the plant, while the other part functions as a lockup storage box, and a dish for serving the dried peppers at the table. The bottom part, which serves as a pot for the plant, is made of waxed cardboard, and folded gapless to avoid water leaks. This part contains the mixture of vermiculite soil and pepper seeds. The other part of the box becomes a lockup tetrahedral box when closed, which can be used for storing and serving, or as a gift box. Most parts of the packaging get a new function during use, thus generating less waste after purchasing it.

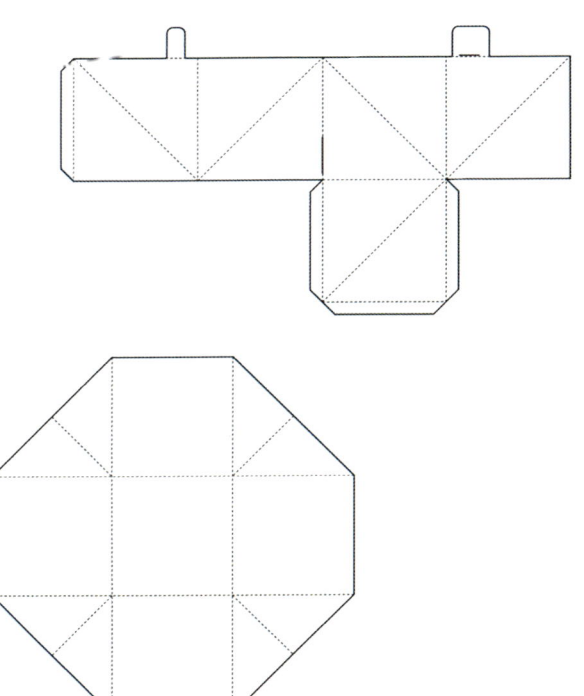

This is a souvenir-like object, that nourishes an active relationship with its user, giving the packaging a new function during daily use. The box contains every item for growing a local type of chili pepper. After opening the box, the user just has to add water to the mixture of vermiculite and seeds, then nurse the seedling following the attached instruction. Finally, the dried peppers can be stored or given away in the closed box.

MARC JACOBS KISS POP COLLECTION

Design agency:
Established

Creative director:
Sam O'Donahue

Client:
Kendo

Country:
USA

Award:
Diamond Pentaward 2015

What's Unique?

Taking inspiration from the original range of cosmetics for Marc Jacobs Beauty which was also designed by Established, these additional products for the line reference crayons and paint sticks, inventing new and fun ways to put colour on your face.

Established designed a collection of fun and playful products for Marc Jacobs Beauty. The line includes Twinkle Pop EyeStick, Kiss Pop Colour Stick and Smart Wand Tinted Face Stick.

MONTEZUMA, COMPOSITE PYRAMIDAL PACKAGING

Designer:
Marija Brkic

Photography:
Teodora Nenadov

Country:
Serbia

Award:
Special award by "Nomen Est Omen"

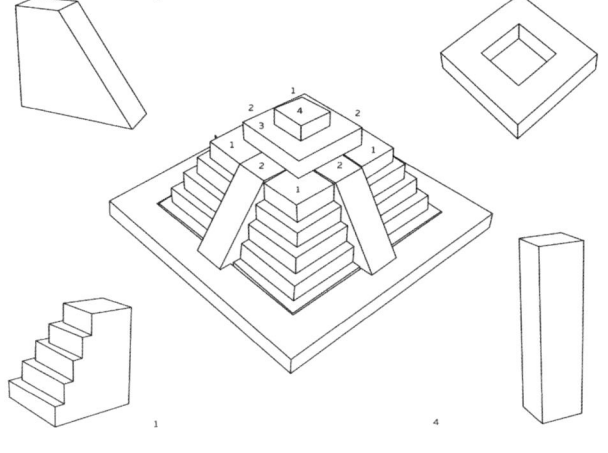

What's Unique?

The project Montezuma is based on the research in the packaging with prolonged didactic influence after its primary usage. Inspired by the ancient civilizations of Mayans and Aztecs, this composite packaging has an educational role. The pyramidal packaging includes ten different shapes which carry chocolates. All the parts are illustrated with different motives in a manner of the ancient fresco-painting.

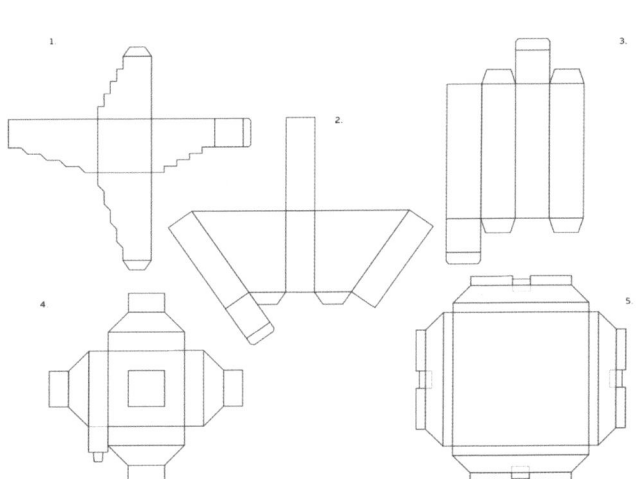

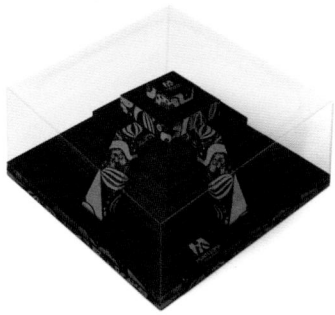

This work includes a child psychology, which is of crucial importance for performing didactic influence. Through the game and the consumption of chocolate, kids can learn about different geometric shapes, history, architecture, art and mythology of the ancient civilizations of Central America, the history of chocolate and the consumption of cocoa which is recorded inside the packaging.

ORIGAMI PACKAGING

Creative director:
Tomás Velásquez

Designer:
Eduardo Parás, Constanza Justiniano

Photography:
Yerko Lucic

Country:
Chile

Award:
Chilean Design Biennale 2017/ Packaging Design Category

What's Unique?

This packaging isn't just unique in his design. It also has a self locking assemble that is hidden in the diagonal creases of the packaging. A simple solution that makes the production go directly from the die, to the final assembly of the case.

The designers took the concept "hapticity" and developed solutions that could generate visual surprise and desire of manipulation, in order to trigger the emotional aspect of design. Graphically, the solution collects the symbolic aspects of copper, which represents Chile, blending it with a geometrical abstraction of snow.

SLOW

Design agency:
Opus B Brand Design

Creative director:
Paweł Frej

Designer:
Piotr Wiśniewski, Aleksandra Wiśniewska

Country:
Poland

Award:
Bronze The Dieline Awards 2017
Gold Award Creativity International Award 2017

What's Unique?

The package resembles an hourglass – the symbol of time. By twisting the ring in the middle, we obtain two separate, sealed bottles. The bottle design allows for slow dosing of the drink to the glass. Irritating? No. This is the time saved for you. So share your SLOW, and give what is most valuable: time.

After 15 minute

SLOW: Manifesto of peace and balance

Slow is a concept incorporated into the Slow Life culture. It is a soothing drink for those who find it hard to find a moment for themselves in the everyday busy lives. Slow provokes a new ritual of product consumption. Not only does the content of the drink make us slow down but also the packaging creates a pretext for relaxation and reflection.

LANGUAGE PLAY.
LANGUAGE LEARNING.

Design agency:
Amelung Design GmbH

Creative director:
Jonathan Sven Amelung

Illustrator:
Angela Wittchen

Country:
Germany

Award:
WPO World Packaging Organisation
European Design Award
Red Dot Design Award
Clio Award

What's Unique?

Each language game is represented by a stereotyped character with attributes typical of the country of origin. When pushing the slip lid of the game boxes upwards and downwards, the facial expressions of the characters change in a playful manner, as their mouths become longer or shorter.

The concept of the Lingua Simplex educational language learning game follows the premise that all language is communication. As a part of the design, this claim was visualised through a packaging concept that attracts both adults and children. Designed to be appealing and humorous, it reflects the basic character of the Lingua Simplex game, which is based on the principle of paired game cards. The pairs game consists of one card with explanatory symbols and one text card with verbs. The text cards with verbs are read out to the game partner, imitating the illustrated character printed on the packaging.

JAPANESE SAKE KOI

Design agency:
BULLET Inc.

Creative director:
Aya Codama

Designer:
Aya Codama

Photography:
Yoshiyuki Watanabe

Country:
Japan

What's Unique?

The package expresses the beauty of koi, with red patterns directly printed on white bottle which resembles the shape of koi. By cutting the box in a koi silhouette, it visually emphasises the image of koi. This vividly designed package will not only stand out in the store, but also be an art piece to decorate your home.

The purpose for this project was "to create an impressively crafted sake that represents Japan". The designer chose the famous, ornamental Japanese koi fish as the motif. Called a living jewel, koi has beautiful red patterns on a white body, and it's attracting many fans worldwide.

ORANGINA SUMMER LABEL

Creative director:
Kiyono Morita

Designer:
Yuko Takagi

Client:
Suntory

Photography:
foton

Country:
Japan

What's Unique?

Orangina designed the label to resemble a bikini worn during summer vacation in France. To make it realistic, details such as wrinkles and buttons were thoughtfully designed. The designers want the customers to think that they are enjoying a fresh orange on the beach.

All the juicy goodness of fresh oranges, bottled for the hot summer vacation. This is Orangina's limited summer label, launched on Marine Day in Japan. Summer vacation was the origin of the brand concept.

PIQUENTUM ST
VITAL 12/13/14
COLLECTION

Design agency:
Studio Sonda

Creative director:
Sean Poropat, Jelena Fiskus

Client:
Vinski podrum Buzet

Photography:
Sean Poropat

Country:
Croatia

Award:
Eurobest Award
London International Award Red Dot
Communication Design Award

What's Unique?

Vintage year is always denoted on the bottles, but how many of us really understand its meaning? It actually witnesses the natural conditions in which the wine matured; it is its own BIOGRAPHY. But with the use of pesticides, the year shown on the bottle has eventually lost its meaning and flavours are becoming unrelated to the actual weather conditions. The designers decided to rise awareness about the importance of understanding the year indicated on labels and then comparing them. In cooperation with the Meteorological service, the data on weather conditions in the territory of vineyards were collected, and the amount of precipitation shows graphically how much weather actually affects the diversity of nature every year. Circles, as standard meteorological rain symbols, show the amount of rainfall in a particular month. The result are labels that give us the possibility to easily compare the different vintage years.

Labels for St. Vital wine - series 2012/2013/2014 are designed to speak out about the importance of understanding Nature. Each one is different, but together, they pass on immediately how different weather conditions create different tastes inside the bottle.

RAIMAIJON SUGARCANE

Design agency:
Prompt Design

Client:
Raimaijon Pasteurized Sugarcane Juice

Country:
Thailand

Award:
iF Design Award Winner 2018

What's Unique?

The packaging is designed to simulate the look, feel and texture of the sugarcane flavour contained inside. Also, the bottle shape and size have been designed to allow the bottle to snap fit and stack on top of one another. When looking from a distance this design is wholly distinctive on the shelf and easy recognition for the customers.

Prompt Design has collaborated with CORdesign Studio to create a new packaging for Thai Sugarcane Industry. This packaging design is to give sugarcane lovers a whole new experience through a literalist graphic design.

ST.STEPHEN HAIR ACCESSORIES

Design agency:
Prompt Design

Creative director:
Somchana Kangwarnjit

Country:
Thailand

Awards:
Silver Pentaward Winner Cat. Body

What's Unique?

The principal of this collection set is to use packaging design and simple arrangement to appeal to different segments of consumers. Matching the functionalities of bobby pins and elastics with design characteristics that resonate with the target market generates a huge impact on the shelf and makes a sale more likely.

In today's haircare accessories market the level of competition is high. For a brand to standout from the crowd, it needs to be unique and able to differentiate itself from the competition. St. Stephen is a brand new hair accessories line carrying products such as bobby pins, elastics, hair pins and head wraps. The most popular products are bobby pins and elastics.

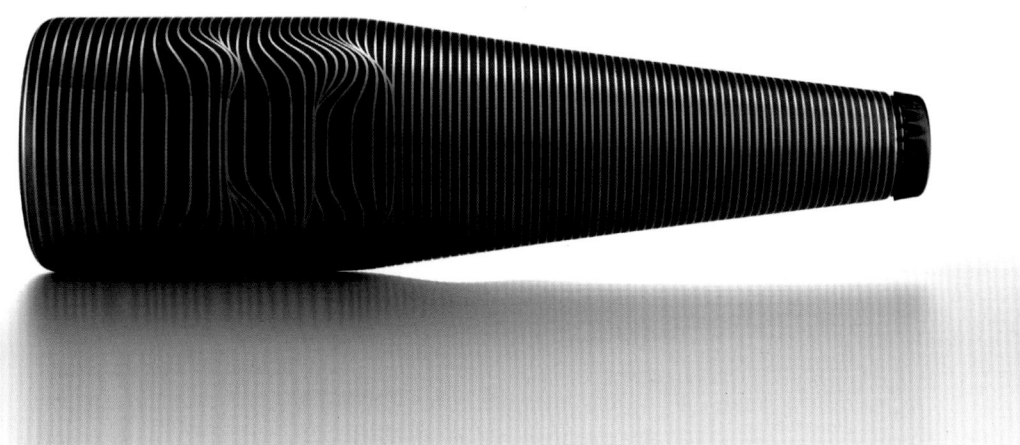

SUB

Design agency:
Reynolds and Reyner

Creative director:
Alexander Andreyev, Artyom Kulik

Client:
Sub Drinks

Country:
Ukraine

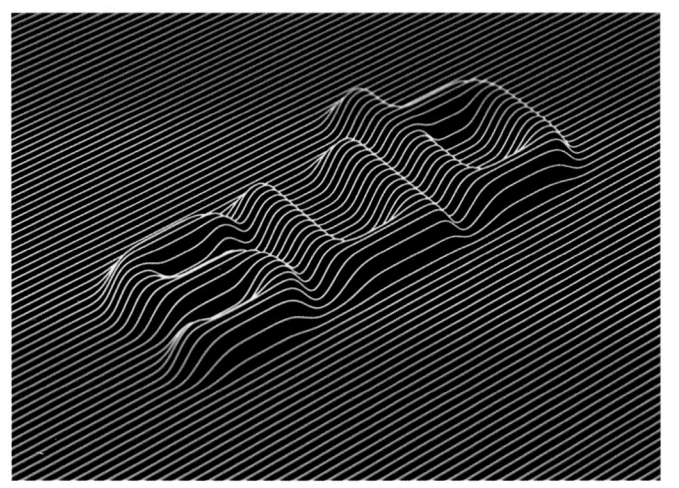

What's Unique?

Sub energy drink is classically canned showing parallel silver stripes on the black background of the bottle surface. The stripes imitate the first brand name letter in 3d format keeping them seamless. Each stripe on Sub packaging symbolises a person's pulse line. The line which appears only when you drink Sub

Energy drink with 3D effect. The idea behind this branding is to find a special unique look able to visualize the effect of the drink avoiding standard branding methods.

CUATRO ALMAS/ BLACKBOARD

Design agency:
Grantipo

Creative director:
Sergio Daniel García

Client:
Bodegas Señorío de Somalo

Country:
Spain

What's Unique?

The blackboard allows consumers to show their affection for their partner by way of illustration or the written word but which can be changed and motified at any time with the swipe of a towel. The bottle, which is painted in a chalkboard-like material, comes equipped with sticks of chalk so that the consumer may personalize it in any way they please. And if they're indecisive, erase their message and start over.

We live in a society that's becoming more spontaneous and indecisive. This idea lead the designers to create the first bottle that permits the consumers to graphically express their feelings.

1872 SHOE PACKAGING

Creative director:
Maribeth Kradel-Weitzel

Designer:
Gabi Stahley

Photography:
Gabi Stahley

Country:
USA

What's Unique?

Inspired by the topography that makes up the landscapes of National Parks, these hiking boots sit within its landscape in the box. The box is created out of layers of cardboard tapering in where the shoes lay at the bottom. Below the shoes is the mission statement of 1872 explaining how they support and how much proceeds are donated to the National Parks. The lift off lid is clear with the topography lines printed to replicate the lines of the landscape from an ariel view.

1872 is a company that donates a percentage of their profits to the National Parks program in America. Another product from 1872 is a series of bandanas dedicated to all different National Parks. They are made from a cyanotype print and have the topography of the park on it. Inside the package is a card with 1872's mission statement and a list of ways to use the bandana for survival on the opposite side.

TOOTHPASTE
FOR KIDS

Designer:
Miriam König

Country:
Germany

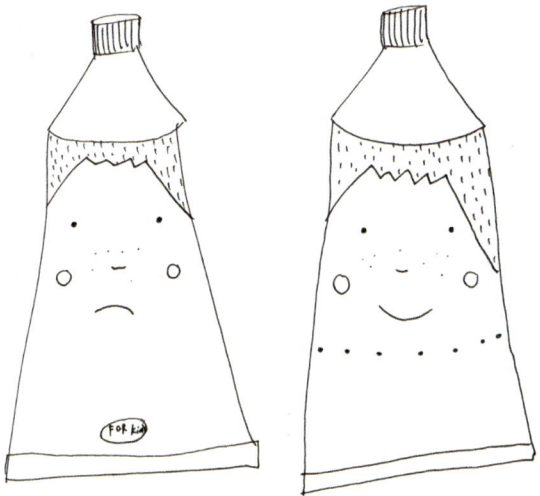

What's Unique?

If the tube is put under water, the package band dissolves, its a special paper that dissolves whengetting wet. Where the former package band was placed now cheerful characters will appear.

The task was to create four different concepts to one product. The challenge included also to think about the different target groups. The main aim was to make teeth-brushing more interesting for children. That's why the designer drew various, colourful characters to make the toothpaste more appealing to children. Characters with black teeth, sweets and toothache are placed on the package band.

OMDESIGN 2016 SELF-PROMOTION PACKAGING

Design agency:
Omdesign

Creative director:
Diogo Gama Rocha

Photography:
Omdesign

Country:
Portugal

Award:
iF Design Award 2018
Red Dot 2017
Pentaward 2017
A' Design Award & Competition 2017

What's Unique?

The top of the packaging is made of wood and its base is a detachable cork agglomerate monoblock, free of any oil derivatives. Inside, under an LBV Vintage Port 2012, there is a real acorn, covered with soil, challenging consumers to grow a cork oak in the packaging base itself, and start a planting cycle.

This eco-packaging marks the launch of the company in 1998 and the 89 awards received in 2016. To create it, Omdesign followed the shape of an acorn, the seed of the company's commitment to the future, while joining 2 Portuguese icons: Port Wine and Cork.

RP10 PACKAGING

Design agency:
Omdesign

Creative director:
Diogo Gama Rocha

Client:
Ramos Pinto

Country:
Portugal

Award:
A' Design Award & Competition 2017
The Dieline Awards 2016
International Design Awards 2016
Graphis Design Annual 2016

What's Unique?

Due to the unique properties of cork, a 100% natural and sustainable material (a very Portuguese product, as well as Port wine), such as lightness, impermeability and thermal insulation, this packaging may be used afterwards for multiple purposes, for example as a cooler, somewhere to keep your pocket contents, pen holder, jar or even a flower vase. "Reuse this packaging in an original and creative way" is the challenge to consumers on the packaging. This ensures that the packaging is not just used to transport and display the wine, but can be reused creatively.

The packaging is 100% Eco-friendly, complying with the environmental and sustainability concerns that Casa Ramos Pinto has made its policy. It consists of two cork blocks and an outer sleeve of pressed cardboard that unites the two parts.

B-ING FLOWER DRINK

Design agency:
Prompt Design

Country:
Thailand

What's Unique?

The 'B-ing' bottle package is designed so that shrink film wrap (which is double-side printed) when being torn along the dotted marks around the bottle top will show a blooming flower appearance with petals.

Beverage market nowadays has been very active in promoting new products and packages. Prompt Design discovers an interesting idea to develop a perception through new packaging design.

The 'B-ing' comes with 4 flavours being the Orchid, Chrysanthemum, Lotus and Butterfly pea. This new package will certainly give customers' experience and feeling of fresh and natural flower flavour.

HOLCIM AGROCAL

Design agency:
Studio Sonda

Creative director:
Sean Poropat, Jelena Fiskus

Designer:
Andrej Glavicic

Illustration:
Eugen Slavik

Country:
Croatia

Award:
Red Dot Communication Design Award
The Dieline Awards

What's Unique?

The solution aims to wrap the product in ecological paper, easily degradable, packaging that uses only one colour in print. It is set in a wooden box that is not supposed to be disposed of, but repurposed and reused by gardeners, for example, for seedlings, storage tools or the like. Different motivational sayings are carved on the box in order to encourage gardeners to a kind of collecting.

In devising the solution, the principle of functionality within the retail space was the key point: the packaging is designed in a way that the boxes can be stacked one upon the other, becoming an independent entity without having to struggle for a place on the shelves in the usually crowded agriculture supplies stores. Another very important functional element for the customer is the possibility to have an easy transport.

At the same time the product is visually well embodied aiming to stand out with its neatness and aesthetic delicacy among the other products within the store.

Agrocal powder is a completely natural and environmentally friendly source of calcium and magnesium for a rapid and effective liming. After the large packaging intended for crops, Holcim has decided to offer the product to urban gardeners as well, in a small package of 4 kg.

Chapter 5
Technology Application in Interactive Packaging Design

I. Application of Material's Sensitivity

There are various materials available for interactive packaging design, which react acutely to external factors such as light, electricity, humidity, temperature, pressure, etc. The designer could achieve the product's artistic expression through these properties. For example, we could use temperature-sensitive ink to print the packaging. When the temperature is lowered to a certain value, the surface of the packaging will change.

See Picture 5.1.1, Picture 5.1.2 Work by Pesign Design. Freely is a limited edition of acidified milk product launched during Brazil Olympic Games. The packaging style embraces Brazilian hot feelings and combines with the air brush style of popular colour run. The printing uses temperature-sensitive ink so the product will show different colours when in cold storage and normal temperature.

II. Application of Material's Reaction Principle

Common reactions such physical changes and chemical changes are also applied in interactive packaging design. For example, some wine packaging uses copper's oxidation

5.1.1

5.1.2

stain to show the time passed. Once the outer copper is completely green, the wine is ready to be enjoyed after sufficient fermentation. To apply these reactions, the designer should know relative knowledge and test the objective factors such as reaction time and quantity through a lot of experiments.

See Picture 5.2.1, Picture 5.2.2 Work by Pavla Chuykina. Gin must be cooled before serving. Prepared in this way gin leaves a very pleasant aftertaste. The bottle is designed into two layers: one for gin, the other for water. Just put the bottle into a fridge and the water in the special tank will turn to iceberg and keep the gin cool and leaving a tasty flavour.

III. Application of QR Code

QR is the abbreviation of Quick Response, which is commonly called two-dimensional code. Large information capacity, high reliability, low cost and strong anti-falsification ability are its obvious advantages. Besides, it can show multiple information such Chinese characters and images. It is very convenient to use. With the common application of smartphones, the application of QR code is increasingly common in interactive packaging design. It can trace product origin, lead users to visit company's website, play music, view pictures and videos. The designer can use content contained in QR code to achieve interaction with users, thus providing further understanding for them.

5.2.1

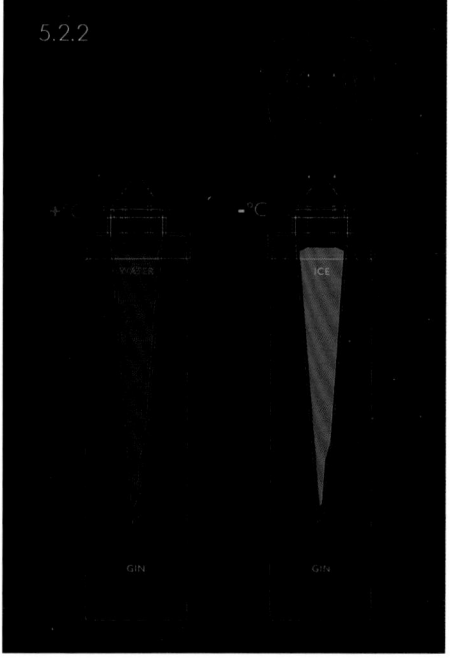

5.2.2

See Picture 5.3.1, Picture 5.3.2 Work by Galima Akhmetzyanova and Pavla Chuykina. This rum is called Ghost Ship, which is inspired by an old story. The ship was built in 1884, in New Zealand. Not long after this she left for North America but disappeared along the way. Someone claimed that the ship got stuck on a reef and sunk but someone claimed to have seen her drifting. The last place the Ghost Ship was spotted was around Auckland, New Zealand. She appeared out of nowhere and then vanished into thin air.

This mysterious story inspired the designers. They have printed a QR code on the bottle and the disappearing Ghost Ship might be seen on a smartphone.

IV. Application of Electric Components

Some high-tech interaction needs electric components to realise. For example, electronic chips are installed in the packaging and NFC, RFID and BLE technologies are used so that users could acquire related product information, safety alert, measurement reminder, expiration monitor, etc. In this way, interaction between product and users are established and it is positive to improve the product safety and logistics tracking.

See Picture 5.4.1 Work by Ong Wongnawa. This design aims to use interactive technology in packaging design to enhance user experience in dealing with dietary supplements. The embedded smart bottle cap acts as a communicator between the package and product to user's smartphone and reminds time of taking pills. Users can also get notification, suggestions for other products based on their recent diet and activity from their wearable tracking device. It can also keep track of inventory and alert users when running low. This convenient interaction provides great convenience for users' daily life.

5.3.1

5.3.2

5.4.1

5.5.1

5.5.2

V. Application of AR Technology

AR is the abbreviation of Augmented Reality, which is a technology that calculates camera image's location and angle and adds corresponding image, video and 3D model. It aims to achieve interaction between virtual world and real world on the screen. After 20 years development, AR technology now is applied in various fields, including health care, military, game entertainment, industrial engineering and video communication. Applied in interactive packaging design, it can satisfy users' experience requirement, which both increase the product's interestingness and ensure interaction between product and users.

See Picture 5.5.1, Picture 5.5.2 Work by Williams Murray Hamm. With its name – Juice Burst, the designers literally burst the fruit via Blippar to emphasise the high constituent of fruit. Consumers can also play games on its app. Nine months after launch, revenue has almost doubled. The juice was the most popular juice brand of the year.

AUKTYON ON THE SUN
(ALBUM COVER)

Design agency:
Great Advertising Group

Creative director:
Andrey Danskov

Country:
Russia

Award:
Red Dot 2017 Winner
Eurobest 2017 Silver
Cannes Lions 2017 Entertainment for Music / Bronze
Red Dot 2017 Best Of The Best Design Packaging
ADC*E 2017 Integration & Innovation (Best Use of Technology) Silver
Clio Music Award 2017

What's Unique?

In order to get the desired result, the designers made dozens of samples, since the photosensitive pigments by different brands in combination with paper and laminate behaved unpredictably, so they needed to find the most reliable option for print. All in all, the designers successfully have found the options and made the idea come true – the print was totally invisible in the shadow but emerged on the sun's rays only.

The packaging was the essential part of the big promo campaign of the album. Fans could listen to release through the app that played tracks only if a smartphone was faced to the sunlight.

Auktyon is an iconic Russian music band. The musicians have recorded their first album "On the sun" – so the designers have created and produced the cover for the new album in such way that image could be seen only if you put a box directly in the sunlight. The idea of the packaging was welcomed and highly awarded by the famous design and advertising prizes like Cannes Lions, Clio Music Awards, and Red Dot. The album is out of stock now because of the popularity, but you can find it in Berlin's Museum of Communication.

FREELY

Design agency:
Pesign Design

Designer:
Peng Chong

Country:
China

What's Unique?

The printing uses temperature sensitive ink so the product will show different colours when in cold storage and normal temperature.

Freely is a limited edition of acidified milk product launched during Brazil Olympic Games. The packaging style embraces Brazilian hot feelings and combines with the air brush style of popular colour run.

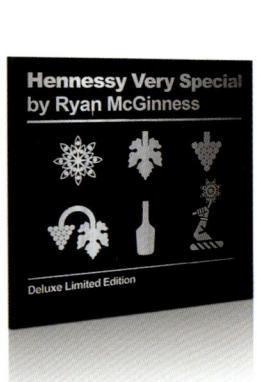

HENNESSY V.S LIMITED EDITION BY RYAN MCGINNESS

Design agency:
www.mediane-creation.com

Artist:
Ryan McGinnness

Brand:
Hennessy

Country:
France

What's Unique?

The limited edition Delux Gift box consists in two offers. One of which is uniquely characterised by a fluorescent label that lights up under the influence of black light. Mediane's mission consisted in accompanying Hennessy with Ryan McGinness's work on all of the developed products : labels, box sets, key visual, goodies...

Hennessy trusts Mediane with the limited edition Ryan McGinness. For its 250th birthday, Hennessy called upon the New Yorker Ryan McGinness to reinvent the famous Hennessy VS bottle.

HOMEBUSH/GIN

Design agency:
BimBom — ideas for fun

Designer:
Galima Akhmetzyanova,
Pavla Chuykina

Photography:
Pavel Gubin

Country:
Russia

What's Unique?

The label is made with a printable luminescent ink which glows in the darkness.

What's going on in the shadows?

When the lights go out, bad things tend to happen. Sounds dramatic, but it's not that bad! In fact, the designers believe that life in the shadows is more diverse. You will never know about this hidden world of thieves, felons and smugglers unless you are a person of adventure.

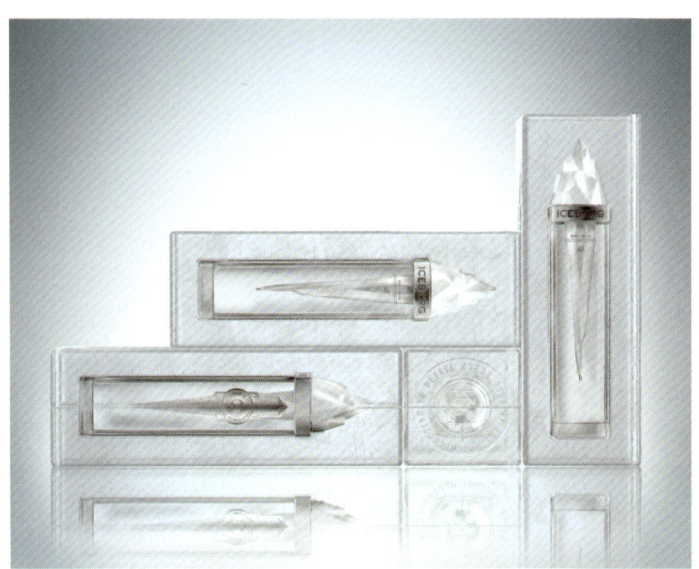

ICEBERG/GIN

Designer:
Pavla Chuykina

Photography:
Maxim Kadashov

Country:
Russia

What's Unique?

The bottle design allows to the drink to be cool much longer due to the ice inside. Just put the bottle into a fridge and the water in the special tank will turn to iceberg and keep the Gin during the party.

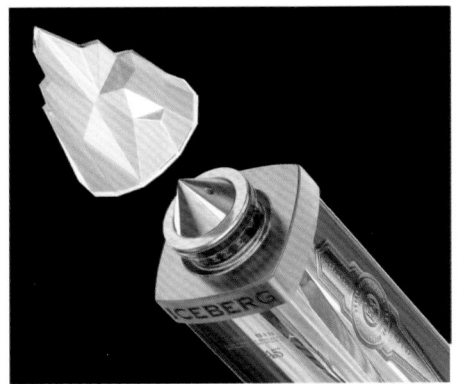

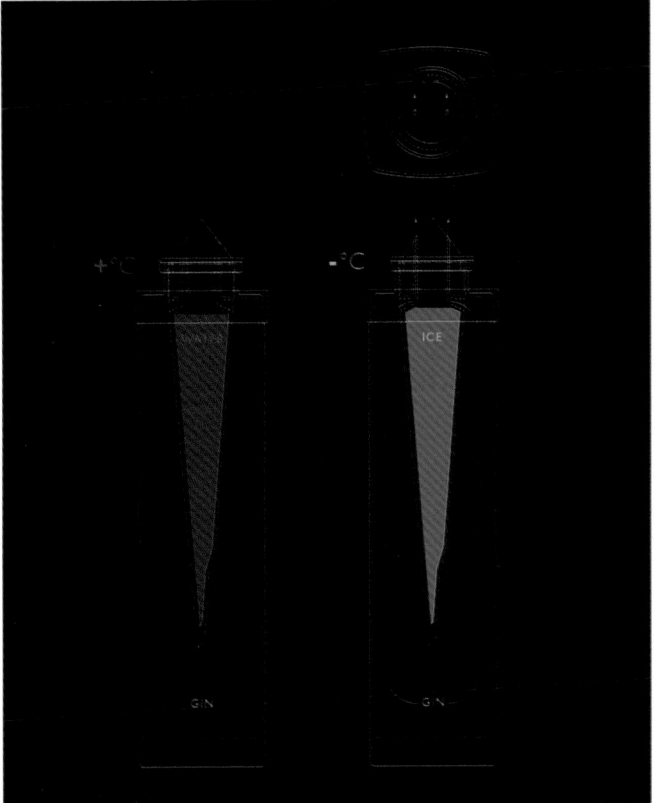

Gin must be cooled before serving. Prepared in this way gin is less scalding and leaves a very pleasant aftertaste.

KARLOVAČKO LEMON NATUR RADLER

Design agency:
Bruketa & Žinić&Grey

Creative director:
Drago Mlakar

Art Director:
Tanja Pruzek Simpovic

Designer:
Tomislav Šestak

Illustrator:
Vedran Klemens

Country:
Croatia

Award:
Directory Magazine

Copywriter:
Vanja Cinic

What's Unique?

The bottle, illustrated with Vis landmarks, was packed in a paper box together with freshly picked lemons from Vis and a special letter whose content could be read only if smeared with lemon juice.

Karlovacko Lemon Natur Radler is the first on the Croatian market made with local lemons from the island of Vis. The packaging and B2B direct mailing tell the story about the 100% natural ingredients and the authentic Vis lemons.

FRESH LABEL

Design agency:
TO-GENKYO

Designer:
Naoki Hirota, Yuki Ijiri, Koji Takahashi

Country:
Japan

What's Unique?

The hourglass-shaped label contains special ink that changes colour based on the amount of ammonia emitted by the meat, the older the meat, the more ammonia it releases. When the meat is no longer suitable for sale, the ink blocks the barcode at the bottom so that it cannot be scanned at the cash register.

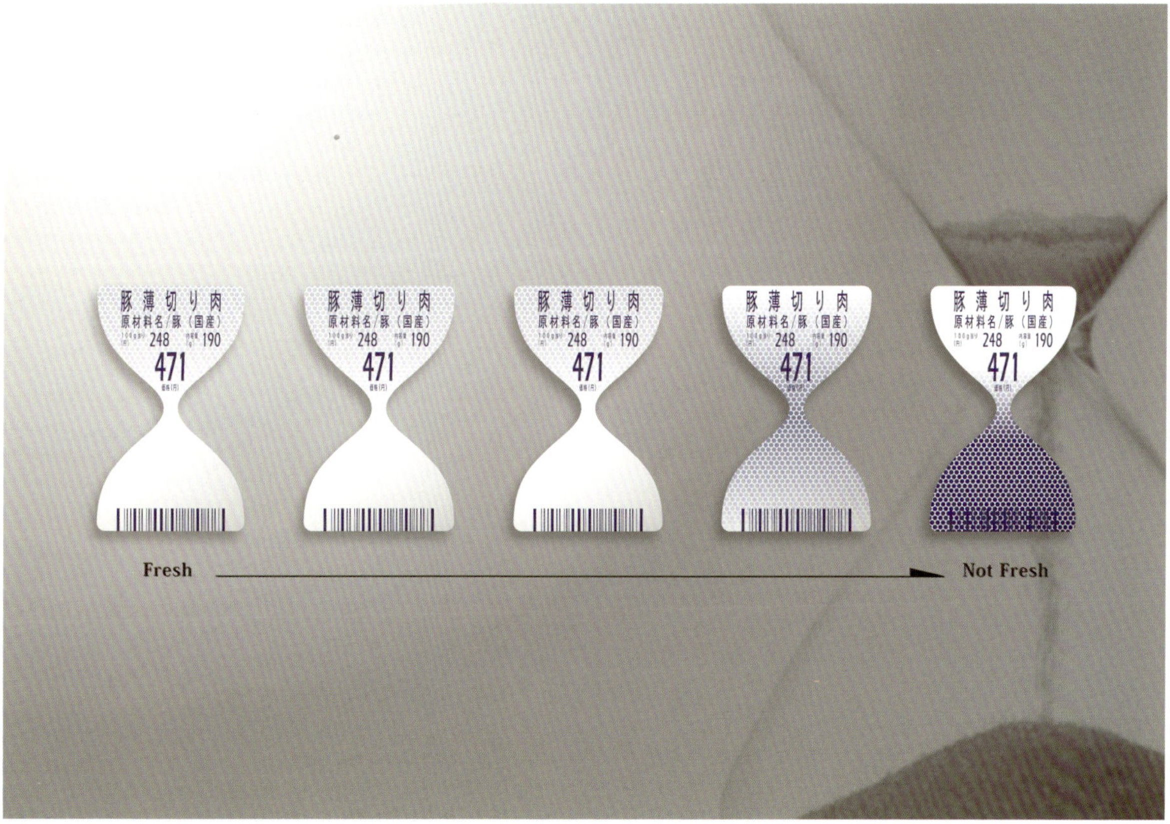

Fresh ———————————————▶ Not Fresh

False labeling on food is a worldwide concern. The proposed food label changes colour in reaction to the ammonia emitted by food as it becomes spoiled. When it is no longer edible, the label darkens to reveal a pattern that renders the barcode unscannable. Designed in the shape of an hourglass, the label instantly conveys its purpose to consumers. This active visualisation of the product's shelf life creates a new relationship between consumers and comestibles.

GEKKO

Design agency:
Opus B Brand Design

Creative director:
Paweł Frej

Designer:
Karolina Starowicz, Paweł Organ, Paweł Frej

Country:
Poland

Award:
Silver Pentaward 2017
Red Dot Design Concept Winner

What's Unique?

The key feature is a system of attaching packaging to shower surfaces which keeps the product out of shower caddies. The system of suction cups integrated with packaging is inspired by the physiognomy of geckos. It allows you to place your shampoo or gel quickly and intuitively in the most convenient part of the shower. Stable fastening and a flexible front of the package facilitate dosing.

Men have been struggling to maintain a place for their cosmetics in showers. "Premium" spots in shower caddies are reserved for their wives and girlfriends. Men are left with the shower floor to put down their shampoos and shower gels. Gekko is a personal care brand concept that will help men to alleviate this problem. Minimalistic design puts the functionality of the product in the foreground. In order to quickly identify basic categories of products, two contrastive colours of packaging have been used – one for shampoos and the other for shower gels. After all, most men buy just the basic variants.

NORR SNO

Design agency:
Opus B Brand Design

Creative director:
Paweł Frej

Designer:
Marcin Górski, Paweł Organ

Country:
Poland

Award:
Gold Epda Award Future Packaging 2017

What's Unique?

The attractive cooling process and an icy mist going through the inside of the bottle create a magic show. Upon turning the base by 90 degrees, the container with gas is opened and the gas escapes to the tunnel leading to the outlet of the package. By transforming from liquid to gas, it derives energy from the surroundings, thus cooling the bottle.

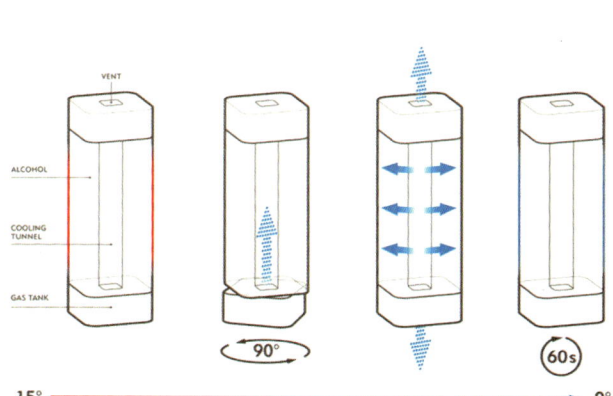

Sno means a cold wind of Scandinavian fjords. Very powerful in its nature, it may even cause small tornadoes. Its immense wind power may quickly reduce temperature by even 20°C. Norr SNO – a self-cooling alcohol brand – has been inspired by this force of nature. Using the phenomenon of physical change of state of gases, it reduces the temperature of alcohol by 15°C within just 60 seconds, regardless of where you are.

FORTUNE WATER

Designer:
Carlos Teles

Country:
USA

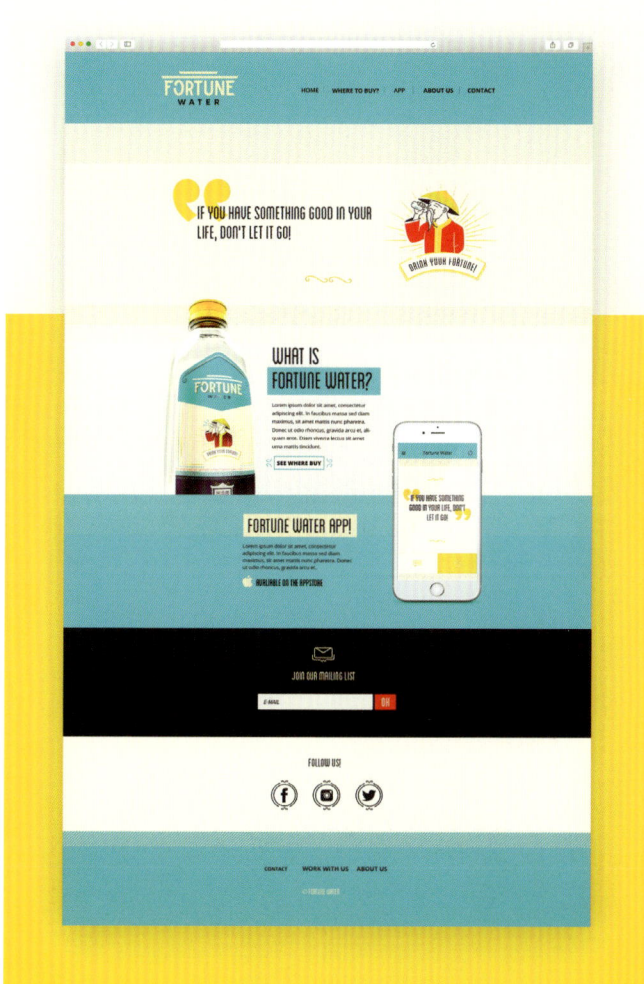

What's Unique?

Through a QR Code printed on the label, any person can scan and visit the fortune water website or app, and there, they can found a cool message with one fortune saying.

Fortune Water is a project for a small company from NY. The idea is to join a water bottle with the fortune sayings, like the fortune cookies. So, all visual identity is inspired in Chinese visual elements, like ornaments, textures, gold details and symbols.

GHOST SHIP/RUM

Design agency:
BimBom — ideas for fun

Designer:
Galima Akhmetzyanova and
Pavla Chuykina

Country:
Russia

Award:
Bronze Pentaward Concept 2014

What's Unique?

The Ghost Ship might be seen on a smartphone — follow this link and see if you can spot her before she disappears.

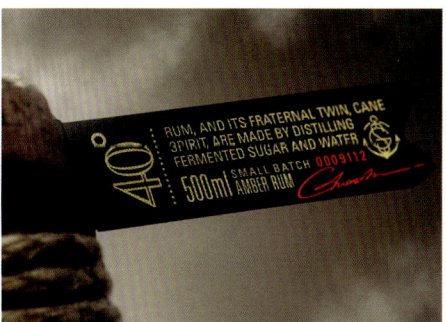

The ship was built in 1884, in New Zealand. Not long after this she left for North America but disappeared along the way. Someone claimed that the ship got stuck on a reef and sunk. However, there have been numerous accounts from eyewitnesses claiming to have seen her drifting. The last place The Ghost Ship was last spotted around Auckland, New Zealand. She appeared out of nowhere and then vanished into thin air.

CUATRO ALMAS/ STEEL

Design agency:
Grantipo

Designer:
Sergio Daniel García

Client:
Bodegas Señorío de Somalo

Country:
Spain

What's Unique?

The designers also created an interactive experience that was integrated within the design of the packaging. By using QR codes printed on the cork and the label, consumers could keep the conversation going. Upon scanning the code on the label they were redirected to a microsite where they could leave a message for other consumers. This message would then be printed on the next bottle up for sale, embedded in the QR code on the cork and thus initiating a new conversation with the opening of every bottle.

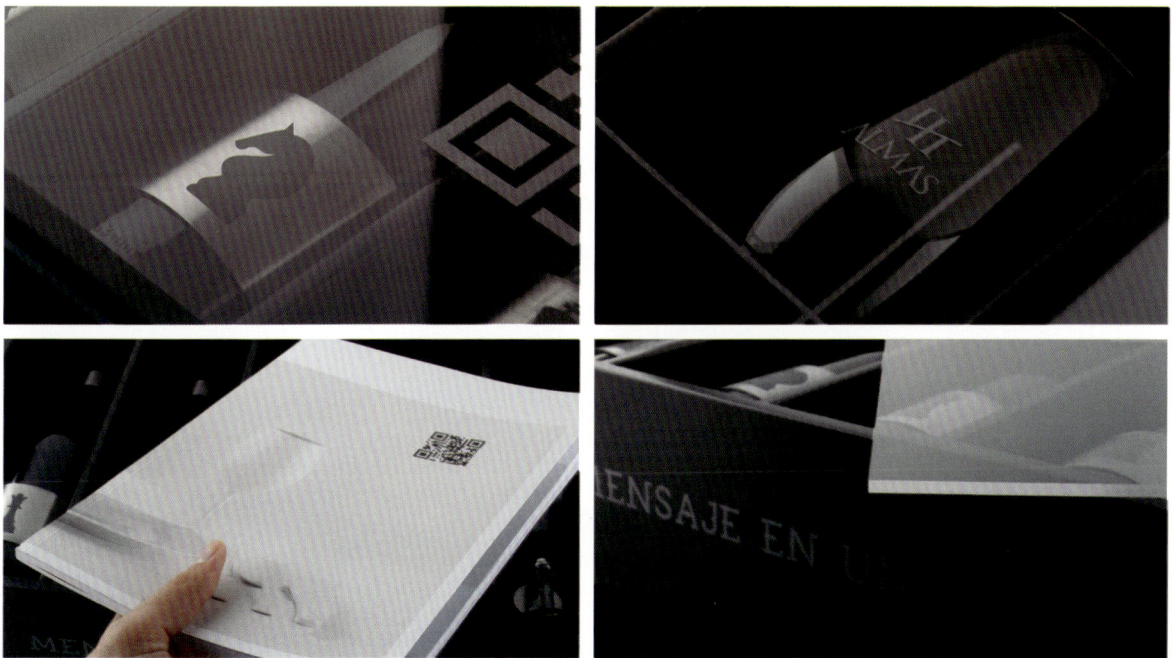

A label made of stainless steel that conserves the wine at a perfect temperature throughout the time it takes to have a conversation (30 to 60 minutes).

If you can't improve a wine any more because its already perfect, the challenge is to keep it that way (between 16°C and 18°C). That's why the designers created a 3.5mm thick stainless steel label that maintains a temperature which helps to conserve the qualities of the wine.

OAK WINE

Designer:
Sergio Daniel García

Client:
La Despensa

Country:
Spain

What's Unique?

A QR code is printed on the bottle, consumers could scan it and get relative information about the wine.

The idea is to create a bottle made with the same wood of cask used to keep the wine, so the designers won't break the cycle of fermentation such as when using other materials. Thus keeping the wine in the same habitat from the beginning to the end of the process.

SYMPHONY OF FLAVOURS - ARTISAN CHOCOLATE

Designer:
Lhente Strydom

Country:
South Africa

Award:
The Loerie Awards 2016 (Bronze)

What's Unique?

Users could scan the QR code through smartphones to enjoy classical symphonies.

The pairing of music and chocolate not only creates a new and unique way of experiencing chocolate, but it will also enhance the flavour of each dark chocolate.

The project involved rebranding the De Villiers Single Origin dark chocolates in a way that showcases the dark chocolates' unique and sensory flavoury palette. In order to communicate the unique flavours of each original chocolate, the flavours were matched with classical symphonies written by South African composers.

MINNAMAME

Design agency:
Yindee Design

Creative director:
Anuchit Panyawatchara

Client:
Lanna Agro Industry Co., Ltd.

Country:
Thailand

Award:
Red Dot Award
TOP Award Asia
Good Design Award
Demark Award

What's Unique?

Minnamame belongs to everyone. Therefore, not only is it the customers' but also the farmers' who grew them. The QR code on the back of the pouch lets consumers discover the farmers and growing place of the Edamame in their very hands. Scan the QR code, our farmers' pride are embodied in every pouch.

Minnamame is everyone's Edamame. The design is half Japanese half Thai. The designers combined Japanese minimalism and a smile of the land of smile. A big bright green brush-stroke smile on white background. No more, no less. Just let the product speak for itself with hospitality and humbleness.

MEKFARTIN BLACK HOLE

Design agency:
Martin Fek

Client:
Mekfartin

Country:
Slovakia

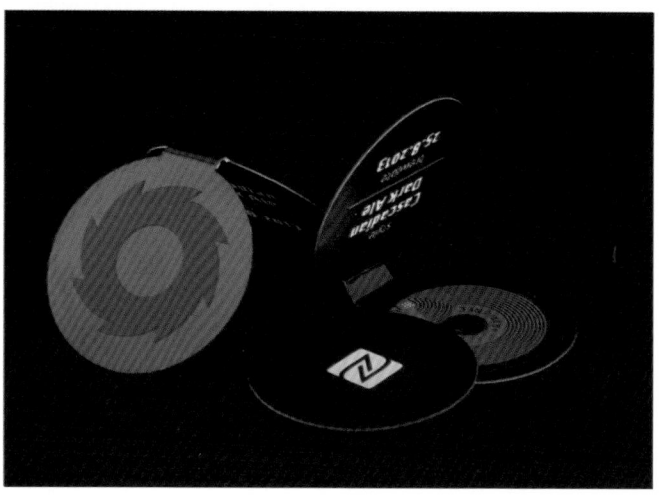

What's Unique?

The neck label contains basic information and additional details in the NFC chip. The users can get these information through smart phones.

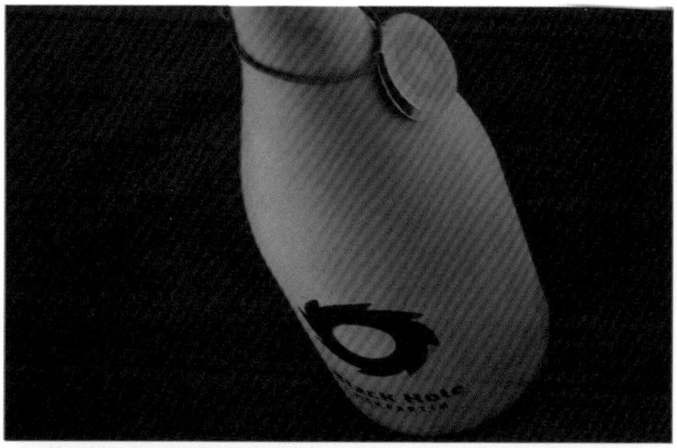

Thick black beer hopped with delicious aromatic Australian hop with cosmic name Galaxy. In the middle of the galaxy supposedly lives black hole, that absorbs everything. It is the same in this beer equivalent. Black Hole absorbed great characteristics of the IPA beer style and tastes of dark roasted malts.

NUTRILINX DIETARY SUPPLEMENTS

Instructor:
Gerardo Herrera

Designer:
Ong Wongnawa

Country:
USA

What's Unique?

The embedded smart bottle cap helps user keep track of their supplement pills ; acting as a communicator between the package and product to user's smartphone. Users can also get notification to remind, suggestions for other products based on their recent diet and activity from their wearable tracking device. It can also keep track of inventory and alert users when stock is running low. This seamless interaction enhances and simplifies the process of taking and managing dietary supplements.

This conceptual design project explores how interactive technology in packaging design can enhance user experience in dealing with dietary supplements; which is a complex food and health product.

JUICE BURST

Design agency:
Williams Murray Hamm

Designer:
Grant Willis, Rachel Price

3D Illustrator:
Jason Budgen

Country:
UK

Award:
DBA Design Effectiveness Gold Award 2016

What's Unique?

Juice Burst has been acclaimed as the world's first digitally interactive soft drink winning a DBA Design Effectiveness Gold Award 2016. True to the name, the designers literally burst the fruit to create a compelling brand journey that jumps from packs into the digital world via Blippar, a category first. Nine months after launch, revenues have almost doubled.

The idea was to give consumers a healthy juice drink but with a fizzy drink attitude. Reframing the brand as a 'healthy Tango', WMH made the strategy sing with an integrated, interactive branding approach.